Advance Praise for *Do Hard Things*

"In *Do Hard Things*, Steve Magness beautifully and persuasively reimagines our understanding of toughness. This is a must read for parents and coaches and anyone else looking to prepare for life's biggest challenges."

—Malcolm Gladwell, *New York Times* bestselling author of *Outliers* and *Talking to Strangers* and host of the *Revisionist History* podcast

"Steve delivers a critical message for our current age of posing and performance: real toughness is not about callous bravado, but instead about the ability to navigate difficulty with grace and an unwavering focus on what matters."

—Cal Newport, *New York Times* bestselling author of *Deep Work* and *Digital Minimalism*

"A thoughtful examination of what it really means to have the right stuff."

—Adam Grant, #1 *New York Times* bestselling author of *Think Again* and host of the TED podcast *WorkLife*

"Steve Magness is one of the giants of modern thinking about high performance across domains, blending a broad knowledge of cutting-edge psychology with hard-earned practical experience from the world-class athletes and other experts he coaches. In his new book, he takes on an age-old question—who triumphs, and why, when the going gets tough?—and reveals that many of our cherished instincts and assumptions are wrong. A crucial read for anyone who cares about delivering their best when the stakes are highest."

—Alex Hutchinson, *New York Times* bestselling author of *Endure*

"We celebrate stories of coaches and leaders who practice the 'weed-out' school of toughness, but research shows that is precisely the wrong way to cultivate fortitude. It is past time to bring the stories in line with the science, and that's what Steve Magness does in *Do Hard Things*."

—David Epstein, *New York Times* bestselling author of *Range* and *The Sports Gene*

"In *Do Hard Things*, Steve Magness dismantles the widely endorsed but damaging suggestion that toughness is about bulldozing your way through difficult situations. Magness's version of toughness—'real toughness'—is more nuanced, forgiving, flexible, and learnable. Real toughness means processing stressors thoughtfully, deliberately, and with vulnerability, rather than superficially and rigidly. *Do Hard Things* changed how I think about stoicism and strength, both on the sports field and more broadly, and I can't recommend it highly enough."

—Adam Alter, professor of marketing and psychology at New York University's Stern School of Business and *New York Times* bestselling author of *Irresistible* and *Drunk Tank Pink*

"*Do Hard Things* will change your mind about what it means to be tough. Steve Magness makes a beautiful and compelling case for the value of inner strength over outer strength and humility over bluster. A must read!"

—Annie Duke, author of *Thinking in Bets*

"Steve Magness possesses an incredible range of wisdom and knowledge about the science, psychology, and practical sides of sport performance. *Do Hard Things* is a master class in how to develop resilience, persistence, and confidence under pressure."

—Christie Aschwanden, *New York Times* bestselling author of *Good to Go*

"A must-read book on a timely and timeless topic, written by the perfect person to explore what it *actually* means to be tough. Steve's been thinking about these issues for years, and this book presents a fascinating and, more importantly, extremely helpful new perspective on toughness and how to build it."

—Brad Stulberg, bestselling author of
The Practice of Groundedness

"Steve Magness has established himself as a leading voice in performance optimization and achieving one's personal bestness—arete, as the Greeks say. In *Do Hard Things*, Magness questions long-standing beliefs that toughness is developed through hubris and infallibility. What he reveals is both hopeful and reassuring. *Do Hard Things* is essential reading for anyone looking to cultivate inner strength in a genuine and authentic way."

—Dean Karnazes, ultramarathoner and
New York Times bestselling author

"*Do Hard Things* is an incredibly deep and completely new approach that examines why and how people overcome the toughest situations. Explaining different stories in a very entertaining lecture for the readers, Steve Magness, one of the most recognized authors and thinkers in sports science, gives us a master class on how to develop resilience and skills to perform at our best in difficult situations."

—Kilian Jornet, author of *Above the Clouds*

ALSO BY STEVE MAGNESS

The Science of Running
Peak Performance with Brad Stulberg
The Passion Paradox with Brad Stulberg

DO HARD THINGS

Why We Get Resilience Wrong and the Surprising Science of Real Toughness

Steve Magness

HarperOne
An Imprint of HarperCollinsPublishers

DO HARD THINGS. Copyright © 2022 by Stephen Magness. All rights reserved. Printed in the United States of America. No part of this book may be used or reproduced in any manner whatsoever without written permission except in the case of brief quotations embodied in critical articles and reviews. For information, address HarperCollins Publishers, 195 Broadway, New York, NY 10007.

HarperCollins books may be purchased for educational, business, or sales promotional use. For information, please email the Special Markets Department at SPsales@harpercollins.com.

FIRST EDITION

Designed by Bonni Leon-Berman

Library of Congress Cataloging-in-Publication Data has been applied for.

ISBN 978-0-06-309861-9
ISBN 978-0-06-327587-4 (ANZ)

22 23 24 25 26   LSC   10 9 8 7 6 5 4 3 2 1

For Hillary, who continually shows me what love and compassion are. You are the most genuine, kindhearted person. Your dedication to all those around you is a constant inspiration. I love you.

In loving memory of Tom Abbey, who taught me humility, open-mindedness, and relentless optimism.
In loving memory of Matt Cobb, who taught me what it means to dedicate yourself to exploring your limits.

CONTENTS

INTRODUCTION: HOW WE GOT TOUGHNESS WRONG AND WHERE TO GO

Chapter 1   From Tough Coaches, Tough Parents, and Tough Guys to Finding Real Inner Strength 3

Chapter 2   Sink or Swim: How We Took the Wrong Lesson from the Military 23

THE FIRST PILLAR OF TOUGHNESS: DITCH THE FACADE, EMBRACE REALITY

Chapter 3   Accept What You Are Capable Of 41

Chapter 4   True Confidence Is Quiet; Insecurity Is Loud. 59

Chapter 5   Know When to Hold 'Em and When to Fold 'Em   85

THE SECOND PILLAR OF TOUGHNESS: LISTEN TO YOUR BODY

Chapter 6   Your Emotions Are Messengers, Not Dictators. . . . 115

Chapter 7   Own the Voice in Your Head 139

THE THIRD PILLAR OF TOUGHNESS: RESPOND INSTEAD OF REACT

Chapter 8   Keep Your Mind Steady 163

Chapter 9   Turn the Dial So You Don't Spiral 193

THE FOURTH PILLAR OF TOUGHNESS: TRANSCEND DISCOMFORT

Chapter 10   Build the Foundation to Do Hard Things 229

Chapter 11   Find Meaning in Discomfort 253

Acknowledgments 269

Notes 273

Index 293

INTRODUCTION

HOW WE GOT TOUGHNESS WRONG AND WHERE TO GO

CHAPTER 1

From Tough Coaches, Tough Parents, and Tough Guys to Finding Real Inner Strength

Hard-nosed. Gritty. Playing through pain. Stoic. Exhibiting emotional fortitude. Showing no signs of distress. Persevering. When college students were asked to describe what it meant to be tough, these words and phrases came to mind. Among 160 elite athletes, *perseverance* came out on top. For most of us, as we read these descriptors, a particular image arises. Perhaps it's a football player popping his dislocated shoulder back into place and demanding to be put back into the game, or maybe it's Craig MacTavish, who retired in 1997 as the last player in the NHL to play without a helmet. For others, the image might be a wounded military hero or a mother fighting through discomfort to care for her child. Chances are that visions of individuals overcoming adversity and some sort of pain or suffering lead the way. That's how we traditionally view toughness: overcoming obstacles with a combination of perseverance, discipline, and stoicism. And if we're honest, when the word *toughness* is mentioned, many of us picture a strong brute of a man.

In a coaching career that spanned five decades and three universities, Bobby Knight amassed an impressive résumé. He won more than nine hundred games, the third most all-time in college basketball; reached the Final Four five times; and took home three NCAA national championships. Of all his successes, his 1976 Indiana basketball team stands out. They won every game they played, sweeping through the NCAA tournament with a win over Michigan to seal the perfect season and Knight's first national championship. In the decades since, no team has been able to match their record. Looking back years later, Knight described what set them apart: "That was a team that was almost impossible to beat, because of its toughness, its strength, its size."

For a man who began his coaching career at West Point, toughness seemed easy to define: "Being able to overcome obstacles. You can't feel sorry for yourself." And for the most part, his teams lived up to the definition, playing disciplined, hard-nosed basketball. While the tendency in basketball is to focus on the glamorous, scoring, this team focused on the unglamorous, defense. They pioneered a pressing man-to-man defense that tested its players' discipline, work ethic, and perseverance. And it worked.

There was just one problem. Not everyone was thriving. The man who led his players to such heights is known as much for his winning ways as for his tantrums and abusive antics. Bobby Knight's success on the court is undeniable. His insistence on toughness being a critical factor in performance is similarly backed by both research and experience, but his methods to achieve it are questionable at best, and downright abusive at worst.

There were the tampons hanging from the lockers of players he thought were "soft." There was the frequent cussing out of players and the accusation that he ordered managers to tape pictures of women's genitalia in players' rooms. There was the 1991 tape of one of his tirades: "This is absolute fucking bullshit. Now I'll fucking run your ass right into the ground. . . . I had to sit around for a fucking year with an 8–10 record in this fucking league, and I mean you will not put me in that fucking position again or you will goddamn pay for it like you can't fucking believe." And the time he brought a piece of toilet paper from the bathroom covered in shit to show his players what he thought of them. Then there was the physical abuse marked by the infamous video of Knight choking a player at practice. All in the name of creating Knight's version of toughness.

"Soft." Female genitalia. Questioning manhood. All actions that clue us in on Knight's actual definition of toughness, one founded on showing no weakness, bulldozing through obstacles, and utilizing fear to establish authority and control. A version we would now call old-school in an attempt to place distance between such barbaric practices and the present. But it's an idea that still dominates the playing fields and performance halls of our present. We have a fundamental misunderstanding of what toughness is. And it pervades far more than the basketball courts.

Tough Parenting

Very demanding. Cold and non-nurturing. Controlling. One-way communication. Using harsh punishment. No, we aren't describing Bobby Knight's coaching handbook, but one of the four main parenting styles.

In the 1960s, developmental psychologist Diana Baumrind pioneered our understanding of parenting. Through research and observation, she discovered that parenting styles can be classified based on two factors: responsiveness and demandingness. Baumrind defined responsiveness as "the extent to which parents intentionally foster individuality, self-regulation, and self-assertion by being attuned, supportive, and acquiescent to children's special needs and demands." In other words, how well do parents respond to and meet the needs of their children? After they lose a soccer match, do you greet your child with warmth and support? Or do you go straight into criticizing their play?

Demandingness, on the other hand, refers to "the claims parents make on children to become integrated into the family whole, by their maturity demands, supervision, disciplinary efforts and willingness to confront the child who disobeys." In other words, how high are the parents' expectations for their child, and how much control do they exert to regulate or influence their child?

Plotting these two characteristics, Baumrind found that most parents fell into three categories that lined up with Goldilocks's search for the perfect bed. If a parent was low in demandingness and high in responsiveness, they were too soft, a permissive parent who would let their child get away with just about anything. If a parent was high in demandingness and low in responsiveness, they were too hard, an authoritarian who relied on harsh discipline, with little attention to the child's needs.

Parents who use an authoritarian style don't trust their children to make good decisions. The parent is in charge, and the child is to obey. Authoritarian parents rely on fear, threats, and punishment to ensure that their children make good choices. A typical refrain from an authoritarian parent might be, "You need to do (this) because I said so." In one study of over one thousand parents, only 31 percent of authoritarian parents said that they should "love their child unconditionally." When it comes to motivation, it's about the sticks, not the carrots.

It's easy to identify the authoritarian parent. Upon seeing their child miss a shot, they are the ones who jump straight to criticism. They are the parent who grounds their kid after every subpar test score, locking their child in their room to study without offering support for how to improve her grades besides a trite suggestion to "work harder." The father who perceives his role is to "toughen up" their boys. Commanding them to suck it up, don't cry, grow up, and never show fear. Even as views have shifted, many parents see harsh discipline as not only beneficial but the lack of it as a sign of the "softening" of America. In one study, 81 percent of Americans thought parents were too soft on their children. It's not just coaches; many parents hold on to this idea that too much warmth or support is "weak."

It's not that punishment or expectations are bad things. It's that, one, punishment or telling a child to simply "work harder" doesn't always get results, and two, when the demandingness far outweighs providing support, we end up with an authoritarian parenting style. The "just right" Goldilocks fit occurs when expectations are high, but so is support. High demand accompanied by warmth and understanding. All parents find themselves somewhere on this continuum, and we shift up and down it based on the context. But it's when there is an extreme mismatch between demand and support that problems arise.

While Baumrind's work originally applied to parent and child, the same principles hold with how we treat one another. Somewhere along the way, we've become very confused about what actual toughness is. From coaching to parenting to leading in the workplace, we've taken the demanding part of the equation and forgotten the other side: warmth, care, and responsiveness to others' needs.

To Be Callous

Callous: to harden, to make insensitive, to develop a thick skin. Look no further than the language often used with toughness. We proclaim teams and individuals as "soft," in need of "hardening up," and implore our teams to "show no signs of weakness." We romanticize the *Karate Kid* narrative—after getting bullied at school, our hero grows stronger and comes back with a vengeance, teaching the bully a lesson. In youth sports, we send our kids to run laps or perform burpees not for some specific training adaptation, but to "toughen them up." In the name of toughness, we rationalize the absurd. In *Until It Hurts*, Mark Hyman visited youth sport clubs across the country and found frequent throwing up after workouts, insult-laden tirades, and more. The justification parents provided for teaching eleven-year-olds to go until they puke? "The stern approach is necessary for children to get in touch with their inner lacrosse warrior."

For too long, our definition of toughness revolved around a belief that the toughest individuals are ones who have thick skin, fear nothing, constrain any emotional reaction, and hide all signs of vulnerability. In other words, they are callous.

Compounding our confusion, we've resorted to tying toughness to masculinity and an ethos of machismo. The mentality to never show weakness, grind it out, play through the pain. Our vocabulary is telling. We tell our sons and daughters to "man up" or, in much cruder terms that are heard on playing fields across the country, "stop being a pussy." Or as the famous line from the movie *A League of Their Own* summarized expectations in sport, "There's no crying in baseball!"

Masculinity is so ingrained in our concept of toughness that if you ask a sampling of individuals about who represents a tough individual, a particular image dominates. More The Rock or Vin Diesel than a small female of similar prowess; brute strength with a large dash of confidence and bravado is how we like our toughest individuals. But as we'll come to see, those who display external signs of machismo are often the "weakest." And women, who research consistently shows quietly handle pain better than their male counterparts, might have had the correct definition of toughness all along—one based on reality, not false confidence and bluster.

Our definition of toughness in the broader world is broken. We've confused it with callousness and machismo, of being manly and stoic. The old model of toughness is represented in the Bobby Knight school of coaching, authoritarian parents, and the callous model of leading. It's the myth of an "inner warrior," one built on the misguided notion that at the heart of being tough is a type of callous demandingness. It's a remnant of a time when military-style drill sergeants—and coaches and parents who thought they were—dictated our view of the concept. Toughness has been hijacked. We've prioritized external displays over true inner strength. And there are consequences.

The Downfall of a Callous View of Toughness

On May 29, 2018, the University of Maryland football team had ten 110-yard sprints for their conditioning workout. By number seven, nineteen-year-old Jordan McNair began to show signs of profound fatigue. According to reports, McNair was bending over at the waist and experiencing cramps. This wasn't ordinary fatigue of a player deciding that he could no longer go on. McNair's body was protesting, at its limit, and screaming for help. Instead of pulling the player from practice, coaches and athletic trainers alike goaded him, yelling to "get him the [expletive] up" and "drag his [expletive] across the field." By the final sprint, video footage shows McNair surrounded by teammates, aiding him through the run's final yards at a near-walking pace. After McNair complained of cramps, it took trainers 34 minutes to have him taken off the field and another 28 minutes to call 911. One hour and 28 minutes elapsed between his final sprint and when the ambulance took McNair to the hospital. McNair died in the hospital two weeks later from heatstroke, thanks in part to a horrid medical response but also to an inability to separate the idea of pushing through the pain and actual danger.

Increasingly over the past decade, we've seen a rash of player deaths and injuries partially from a misguided belief in developing toughness. Rhabdomyolysis (or rhabdo for short) is a once-rare condition where damaged muscle products leak into the bloodstream, putting an unusual demand on the kidneys to process it all. In extreme cases, death can occur. A disease once primarily caused by infections or drug use has transformed into a somewhat common occurrence, thanks to a bevy of cases caused by extreme workouts. Endless push-ups, squats, burpees, and other exercises designed not to improve fitness, but to "test" their athletes. As professor of sports business at Ohio University B. David Ridpath described, the true notion of these workouts is not conditioning: "Taking a cue from a head coach with a desire to either toughen up the current players or weed out a few to open some scholarship slots, the strength coach often will 'condition' these players with a vengeance and a mandate to make them suffer." While we may think we've come a long way in athletic performance, the extreme workout in the name of toughening up is alive and still causing harm.

While death may not occur in classrooms or homes from enacting an authoritarian approach to parenting or leading, research shows lasting psychological consequences. Authoritarian parenting leads to lower independence, more aggressive behavior, and a higher likelihood for substance abuse and risky behaviors. In sport, the controlling, demanding style also fails. On the athletic fields, it's linked to lower grit and an increase in emotional exhaustion, burnout, and fear of failure.

Even in terms of discipline, the area that you would think a demanding style would be successful, it falls short. In one study of over 1,200 parents, authoritarian parenting was linked to a much higher rate of child misbehavior. It even fails in places where it seems a natural fit: the military. In the Israeli military, those who grew up in an authoritarian environment adapted to and coped with the challenges of military life much worse than their peers who grew up in a nurturing environment. The authoritarian style creates the appearance of discipline without actually fostering it.

Somewhat ironically, teaching, parenting, or coaching for this version of "toughness" creates fragile and dependent individuals. What does a child who was taught to follow the rules unquestioningly out of fear do when a parent isn't there to dictate his behavior? What does an adult who was taught to rely on fear for motivation do when left to her own devices in the real world? What does a football player who learns to push himself only when a coach is screaming in his face do when it's him alone on the field? The answer lies in how one young athlete responded when questioned about his experience with punishment in sport: "Coaches use exercise as punishment because they want you to become stronger. . . . It gets in your head and you start thinking, 'I need to do better. I need to work harder because I don't want to be punished.'" This young man didn't want to work harder because he wanted to get better, to win the game, or for some sort of internal reason. He wanted to avoid punishment. That's the message we are sending.

Proclaiming the old-school model as the way to develop toughness is akin to declaring that the best way to teach swimming is to throw every kid into the deep end of the pool. For some, it would work, but for many, it would prove disastrous. There are better ways to ensure everyone learns the skills necessary to be truly tough.

Redefining Toughness

Here's the problem: in trying to toughen through callousness, we've trained ourselves to respond to fear and power. The reason we push through discomfort is because we imagine someone is standing over us yelling, or that if we fail, we will face punishment. We've been conditioned to see the external as more important than the internal, and that putting on a facade of toughness ("I'm not afraid of anything!") is more important than how we handle difficult times. Remove the fear, power, and control, and our "tough" individual is left without the necessary skills to navigate adversity. The old view of toughness gives him a hammer and expects him to bash his way through any problem. But truly being tough isn't the same as being callous.

For far too long, we've confused toughness with something far more sinister. We've made the mistake of Bobby Knight and authoritarian parents: confusing the appearance of strength with possessing it, and confusing being callous with instilling discipline. And the truth is, it's all fake.

Fake toughness is easy to identify. It's Bobby Knight losing control and throwing tantrums in the name of "discipline." It's the appearance of power without substance behind it. It's the idea that toughness is about fighting and ass-kicking. It's the guy picking a fight at your local gym. The anonymous poster acting like a hard-ass on message boards. The bully at school. The executive who masks his insecurity by yelling at his subordinates. The strength coach who works her athletes so hard that they frequently get injured or sick. The person who hates the "other" because that's a lot easier than facing their own pain and suffering. The parent who confuses demandingness for discipline. The coaches who mistake control for respect. And the vast majority of us who have mistaken external signs of strength for inner confidence and drive. We've fallen for a kind of fake toughness that is:

- control- and power-driven,
- developed through fear,
- fueled by insecurity, and
- based on appearance over substance.

Yet, we are in a new era, one in which the emerging science and psychology on overcoming challenges point to a radically different definition of toughness. Regardless of whether it's on the sporting field, in the classroom, or in the boardroom, strength and resilience don't come from blindly powering through adversity or pretending that punishing ourselves yields results. Instead, real toughness is experiencing discomfort or distress, leaning in, paying attention, and creating space to take thoughtful action. It's maintaining a clear head to be able to make the appropriate decision. Toughness is navigating discomfort to make the best decision you can. And research shows that this model of toughness is more effective at getting results than the old one.

Real toughness is much harder than the fake kind. To understand what toughness is, we can look to another successful coach. One who allows his players to be who they are, celebrating "the way they see the world." One who encourages meditation and yoga or shifts from a meeting to playing ring toss if players get heated. According to one star player, "He's never negative, doesn't scream. He finds a way to turn a mistake into a positive."

Pete Carroll isn't a saint; he's a coach. After losing his job as an NFL head coach in the 1990s, Carroll stopped imitating what others did and followed his own path. It may sound like he's an easygoing "player's coach" who is "soft" on his team, but Carroll is one of only three coaches in history to win both an NCAA championship and a Super Bowl. He also believes in toughness.

Carroll wants players who come through when the game is on the line. But instead of relying solely on discipline, he believes toughness comes from somewhere much different: from an inner drive to keep them focused, from embracing challenges and bouncing back if things didn't go their way, from perseverance and passion. Carroll doesn't shy away from making his players do difficult things. He embraces it, with his "always compete" practices. But he recognizes it's his job to give them the skills to handle adversity. "Teaching guys how to feel confident enough to believe in what they've been prepared to do and believing what they can do and they go out there and do it," Carroll relayed to The Bleacher Report.

Carroll is trying to develop real toughness—a kind that replaces control with autonomy, appearance with substance, rigidly pushing forward with flexibility to adapt, motivation from fear with an inner drive, and insecurity with a quiet confidence. It's time to move away from a model based on perceived strength, power, and whatever violent military metaphor of fighting we'd like to use. And lest you think Pete Carroll is merely an aberration, consider Don Shula, Bill Walsh, and Tony Dungy, among the most successful coaches in NFL history. In basketball, consider John Wooden, Dean Smith, Brad Stevens, or Mike D'Antoni, who in discussing his approach to player feedback told me, "We keep it positive here." As Ken Reed, author of *Ego vs. Soul in Sports*, summarized, "For every Lombardi or Bobby Knight you give me, I can give you an equally successful—if not more successful—humanistic coach. . . . Despite the success of Wooden, Shula, Dungy, Stevens, and others, our society has conditioned us to think that autocratic coaches are better coaches; that they win more often. It's a myth. But it's a vicious cycle."

Again, this isn't just theory; it's science-driven. In 2008, researchers out of Eastern Washington set out to explore the relationship between leadership style and the development of toughness. After conducting research on nearly two hundred basketball players and their coaches, they concluded, "The results of this study seem to suggest that the 'keys' to promoting mental toughness do not lie in this autocratic, authoritarian, or oppressive style. It appears to lie, paradoxically, with the coach's ability to produce an environment, which emphasizes trust and inclusion, humility, and service."

Real toughness is about providing the tool set to handle adversity. It's teaching. Fake toughness creates fragility, responding out of fear, suppressing what we feel, and attempting to press onward no matter the situation or demands. Real toughness pushes us to work with our body and mind instead of against them. To face the reality of the situation and what we can do about it, to use feedback as information to guide us, to accept the emotions and thoughts that come into play, and to develop a flexible array of ways to respond to a challenge. Toughness is having the space to make the right choice under discomfort.

Whether discomfort comes in the form of anxiety, fear, pain, uncertainty, or fatigue, navigating through it is what toughness is all about. Not bulldozing or pushing through, but navigating. Sometimes that means going through, around, under, or waiting until it passes. When we frame toughness as a decision to act under discomfort, it allows us to see that toughness is far more than merely having grit or grinding through. We can actively change how we appraise, experience, and respond to discomfort. Each step along the way requires a different skill set and approach. It requires many tools, not just a hammer.

The truth is, this model of toughness, of navigating discomfort instead of bulldozing, isn't new. The military, often seen as the epitome of demanding, macho, extreme toughness, has actually been perfecting this other model of resilience for decades. Like the parents who went all in on being demanding, but forgot how to be responsive, we took the drill sergeant but forgot the training and support.

TOUGHNESS MAXIM

Real toughness is experiencing discomfort or distress, leaning in, paying attention, and creating space to take thoughtful action. It's navigating discomfort to make the best decision you can.

Searching for Real Toughness

There were seven men on the starting line. Six wore the blue and gold of University of California, Berkeley, and then there was me, the lone athlete donning the red and white of the University of Houston. We were lined up to compete in the Don Bowden Mile. In 1957, Bowden, a twenty-year-old Cal economics student at the time, became the first American to break the mythical four-minute-mile barrier. Fifty years later, the seven of us were chasing the same feat.

As the gun went off, I charged off the line, slotting myself into third place, positioning myself right behind the designated rabbits (pacesetters who would drop out halfway through the race). With such a small field, tactics were simple: Get near the front so I could lock my gaze onto the back of the runner ahead of me and turn my brain off. Then with a little over a lap left, I'd reengage, using my mental strength to get through the impending pain and fatigue.

For the first half of any race, the less thinking, the better. It's wasted mental effort. No one wins the race in the first half of an endurance contest; they only lose it. Less thinking meant fewer thoughts about the impending pain or doubts over whether I could sustain the pace the whole way. We all have different ways to cope with the burning quads, the searing lungs, and the uncertainty of knowing whether we'll have enough energy to finish. My strategy, honed over dozens of races, was to zone out and go for a ride until it was time for the real racing to begin. I was storing up my mental energy to combat the surge of discomfort that would rear its head over the final lap.

The first lap flew by in sixty seconds. "Right on pace" was the only thought that crossed my mind. The magic of a four-minute mile was that it required very little math—a nice bonus for the oxygen-deprived brain. Even simple math, like counting four laps, becomes surprisingly difficult under such strain. Four laps, all at or under sixty seconds, and the prize is yours. As we passed the halfway mark, a coach yelled from the infield, "1:59 . . . 2:00." Everything was going according to plan. My mind was calm yet focused. It was almost time to come out of autopilot and see what was left in the tank. To see if I could nail the landing.

Every runner has his or her signature of fatigue, their tell sign in the game of poker we call a race. Some runners develop labored breathing as they gasp for much-needed oxygen. Others have a physical tell: a slight rising of the shoulders, a wildly flailing arm, or a straining face that becomes contorted with pain. Fatigue exposes us, cracking even the most stoic and hardened of racers. Every runner knows their tell, and if you race your competitors enough, you come to know theirs as well. A runner in front of you might start leaning slightly back, letting you know that they are losing control of their core. Or the arms might swing just a touch more vigorously, letting you know that their arms are taking over because their legs are failing them. Fatigue unmasks our breaking points.

In the hundreds of races I'd run up until then, my weak point was always in my legs. That's where the familiar friend of fatigue would announce its presence. My breathing, on the other hand, was always reliable—rhythmic and just on the right side of under control—even as the rest of my body failed. Sometimes I'd use this to my advantage, saying a short word or two mid-race, hoping that my competitor might be fooled into thinking I was doing better than I actually was.

We were 900 meters into the 1,609-meter race when I got the first hint that something was wrong: a sensation in my neck, of tightening and straining, along with a strange breath, an almost high-pitched gasp as if I'd swallowed some water and it went down the wrong tube. My inner focus shattered; the mind I was trying to keep on autopilot jumped into action, as if an alarm had gone off in the cockpit. "What was that? What's wrong? Why am I breathing hard? This is too soon. My legs feel fine. You're done. You have almost half of the race still to run. It's over." My calm inner journey disintegrated.

I tried to combat the internal freak-out and flew through all of the tricks I'd honed over a decade of running: breaking the race into manageable pieces, ignoring the fatigue, pushing through it. I wasn't new to this game. Freak-outs were part of racing. And for a moment, it worked. I put my head down, determined to grind my way through whatever it was I had just experienced. I was tough, after all. That's how I made it this far, I thought. Toughness. The final pacesetter was about to drop out, and I was tucked right behind the top Cal runner. A lap and a half to glory: I could hold on.

Less than 100 meters later, my inner voice screamed, "I can't breathe. What the hell!? I can't breathe!" Every time I tried to get air in, I was greeted with a high-pitched gasp, as if my airway had something lodged in it. I pulled off to the inside of the track, coming to an abrupt halt and throwing my head back as if to open my airways. I collapsed onto my knees. After a few panic-stricken moments, it was as if someone had reached down my throat and removed the blockage. I remember thinking, "What in the hell was that?"

As a runner, I'd always prided myself on my "toughness." In high school, I was notorious for throwing up after just about every race, for pushing to the point of exhaustion. As one of my college coaches, Theresa Fuqua, once told me, "You race with intensity. There's never a question of whether the effort will be there or not; it's just whether or not the body is going to produce what your mind wills it that day." Within a matter of seconds, I'd transformed from being in control of my body and mind to losing it.*

For the next year, I'd search for an explanation for what occurred that day. After dozens of tests ranging from scopes down my throat to echocardiograms to a slew of treadmill and biking tests to exhaustion in some doctor's office, I found a diagnosis. After I'd seen half a dozen specialists from across the country, it was an allergist with a keen eye for research and a knack for solving obscure problems that found the answer. Dr. Stephen Miles diagnosed it as paradoxical vocal cord dysfunction (VCD).

---

* Cal runner David Torrence ran his first sub-four mile in this race. David went on to run in the 2016 Olympics. Tragically, David passed away in 2017. David was one of the nicest guys in the sport. And as I reflect on this race now, it's turned from one of my most tragic memories to one of appreciation. I'm thankful that I got to share the track with a person who would become a friend and influence the world in such a positive way. RIP, David.

The vocal cords sit in the larynx in your throat and play a role in their namesake (i.e., making sounds) as well as in respiration. They open widely when you breathe in and partially close when you breathe out. They also have a third function, protection. They shut to protect the lower airways from any objects that may attempt to pass through. The opening or closing of the vocal cords is controlled almost entirely via reflex. No thinking involved, just open and shut, open and shut. With those who suffer from VCD, this process goes haywire. The flaps of the vocal cords malfunction, shutting when they are supposed to open—essentially blocking the airways on inhalation. The current theory is that the vocal cords become hyperresponsive, ready to shut at any moment, like a trigger-happy guard prepared to defend his post at the slightest hint of danger.

A reflex gone wrong. In severe cases like mine, this leads to the unpleasant sensation of not being able to breathe in. Panic, fear, and anxiety frequently accompany the physical symptoms. And as VCD sufferers experience the trauma of not being able to breathe, the fear only worsens.

So what causes the vocal cords to go haywire? How does such a deeply ingrained process cease to function in the way it does for billions of people daily? The American Thoracic Society cites "strong emotions" and stress as triggers that set the disorder in motion. Other researchers point toward a hyperresponsive larynx and a shift in the nervous system activity, a combination that primes the body to respond to a stressor (be it a psychological one or a physical irritant) by closing off the vocal cords. In my case, a normal "freak-out" moment in racing—that either we work through or causes us to slow down—ended up as a full-blown disaster.

For the next few years while trying to figure out what was going on, the activity that I grew up excelling at, the item that largely defined my sense of self-worth, transformed into something I feared. The path that I knew to succeed—pushing until I puked—backfired, exacerbating the issue. To continue to pursue a sport that I loved, I had to find a new tactic. I had to relax, to keep my breathing, neck, and mind steady and under control, all at the exact moment when discomfort and doubts were at their highest. In many ways, this book started the moment I collapsed. A search for what it means to be tough, for understanding how to control an inner world that often goes haywire. What follows is not only what allowed me to race again, but a process that I soon found could apply far beyond the oval track.

Running will serve as a central thread throughout the book. There are a few reasons for this. As you're now aware, it's an activity I'm intimately familiar with, having raced, coached, and studied. But more importantly, it's a sport where you are alone, in your head, navigating immense levels of discomfort. Running and similar tests of endurance provide the perfect backdrop for studying toughness. If you aren't an endurance athlete, don't despair: we'll reach far beyond sport, learning how the same principles can apply to everything from parenting to handling grief to managing and leading people—be they six or sixty years old. The lessons in this book come partially through experience—in working with elite athletes across the professional sports, as well as executives and entrepreneurs in the workplace—and partially through the latest science spanning the fields of cognitive psychology, neuroscience, and physiology. While sports may provide many examples, these lessons apply far beyond the playing field.

What I've learned on this journey is simple: We have a fundamental misunderstanding of what toughness is. Being tough isn't some special attribute reserved for the talented. It's attainable to all. Most of us are just walking around with the wrong framework. Stuck in the old-school mindset described throughout this chapter.

Yet, we have so many examples of regular, everyday individuals who have immense inner strength. Many of whom you'll meet in this book. Individuals who resist putting on the facade of perfectionism and strength and show us the nuance and complexity of being a human with compassion, grit, and grace. As podcaster Rich Roll told me in summarizing the hundreds of interviews he's conducted, "Everybody goes through shit in their life. Nobody escapes obstacles." If we're going to face obstacles, we might as well figure out the best way to navigate them.

Real toughness isn't just about helping you deal with pain or perform better; it's about making you a healthier, happier human being. By adopting the principles of real toughness, you'll learn how to prepare for, communicate with, respond to, and ultimately transcend discomfort. It'll help you navigate arguments, handle your emotions, and wrestle back control of your life when you are on the brink of burnout.

In the chapters that follow, I'm going to take you through the four pillars of real toughness so that you have the tool kit to navigate whatever obstacle you face.

- **PILLAR 1:** Ditch the Facade, Embrace Reality
- **PILLAR 2:** Listen to Your Body
- **PILLAR 3:** Respond instead of React
- **PILLAR 4:** Transcend Discomfort

But first, let's explore where we went off track. Why do so many of us carry around the same model of toughness that Bobby Knight and authoritarian parents hold on to? In order to move forward, we need to understand why our foundation of toughness is built on a facade.

CHAPTER 2

Sink or Swim: How We Took the Wrong Lesson from the Military

In 1954, Texas A&M was far from the cash-rich and athletic-hungry university that it's known as today. It was the "cow college," a men's-only school stuck in the past. As one student remarked of the time, "The campus looked a little bit like a penitentiary." So when football coach Paul "Bear" Bryant picked up and left the University of Kentucky for Texas A&M, not only was it a surprise, but it provided hope for the university's fledgling football team. When Bryant stepped foot on the A&M campus, he knew he needed to make a change, and it was going to start with preseason camp.

In the summer of 1954, Bryant and his team of nearly one hundred players set out for the small Texas town of Junction, situated 140 miles west of Austin, or more precisely, in the middle of nowhere. The team was excitedly awaiting camp. As senior quarterback Elwood Kettler recalled, "There was supposed to be swimming, nice green grass. I was looking forward to it. . . . I thought it was going to be like a vacation." Bryant had other ideas. He was determined to harden his team, to "separate the quitters from the keepers," and to send the message that change was brewing in College Station. Junction provided the perfect backdrop.

"The facilities were so sorry, that just looking at the place would discourage you," Bryant later wrote. And the fields they practiced on weren't much better. "It wasn't a football field, it wasn't any kind of field," Dennis Goehring recalled years later. A blistering heat wave and one of the worst recorded droughts in the Texas Hill Country wreaked havoc on the small town. Practice was brutal, as Mickey Herskowitz reported for the *Houston Post*: "They had a full-scale scrimmage, the very first thing, and guys were throwing up all over the place."

As camp wore on, player attrition mounted. The newspapers of the day even kept a tally. "Sixth Player Quits Team at Texas A&M," read the headline in the *Washington Post*. Rob Roy Spiller, the bus depot attendant at the time, recalled players desperate to escape the training camp from hell. As the boys approached the depot, Spiller asked, "Where would y'all like to go this morning?" The typical reply, "We don't care. First bus out." By the end of the ten-day camp, by most accounts, between twenty-seven and thirty-five players remained. Nearly seventy players had quit. Gene Stallings bluntly summarized the attrition in Jim Dent's classic book on the subject, *The Junction Boys*: "We went out there in two buses and came back in one."

Bryant would go on to achieve legendary status in college football at the University of Alabama, winning six national championships and becoming one of the most revered coaches in history. Before he left for Alabama, Bryant did just what he said he would. He turned Texas A&M into a national title contender, going 9–0 in 1956. The Junction Boys camp was a central component to that success. It transformed the culture of a downtrodden team, developing a core nucleus of players who would overcome any obstacle. As Bob Easley, a fullback on the 1954 team, put it, "You go through ten days of hell, and you go in as a boy and you come out a man." Survive, and you thrive.

The story of the Junction Boys has become a symbol for coaches and players everywhere. While harsh, the training camp was a rousing success. If you want to get the best out of a team, weed out the weak players and harden the remaining ones. Toughening up individuals was the secret to success. It's a story that's been memorialized in a bestselling book and an ESPN movie. It's a story that we've held on to as the blueprint for creating toughness.

But that Junction Boys team, the thirty-odd players who lasted, how did they fare that season? Their first game was a 9–41 drubbing from Texas Tech. The rest of the season didn't get much better. One win, nine losses. Popular lore often overlooks their abysmal results that season and points to A&M's success two years later, when they finished 9–0 in 1956. The Junction Boys camp has gone down in history as the focal point of the turnaround. But like most things, it's easier to assign attribution after the fact than to know the true cause. Only eight players who survived the camp from hell played on the winning Texas A&M team two years later.

John David Crow, a future Heisman Trophy winner, would serve as the backbone of the undefeated team, leading the team in touchdowns and yards. Crow was part of the Junction Boys team, but as a true freshman, he wasn't allowed to travel with the group to camp. The star quarterback for that same undefeated team? Jim Wright, another freshman who didn't go to Junction. Their All-American tackle, Charlie Krueger? Same story. He stayed at home. Years later, Ed Dudley, a member of the A&M team during the Bryant years, summed it up: "Our freshman [in 1954] won the conference [in 1956]."

The eight players who survived Junction played a large role, but so did another change at A&M. They landed blue-chip prospects. Through a combination of Bryant's skill and the help of bending and breaking recruiting rules, A&M got better talent to complement their core players. In his autobiography, Bryant explained, "That first year was brutal. We could hardly get anybody to come to A&M, and I know some of our alumni went out and paid a few boys." Better talent meant better results, regardless of how that talent was acquired.

While tactics like those employed by Bryant have entered the sporting lore as a way to develop toughness, they were anything but. The camp wasn't about creating tough players. It was about sorting, "separating the wheat from the chaff." And even that seemed to fail. Top recruits, future NFL players, and even a future war hero quit after Bryant's antics. The quitters included All-Southwest Conference players Fred Broussard, who would go on to play in the NFL, and Joe Boring, who switched to baseball and led the Aggies to a conference title.

It's tempting to paint the picture that those who survived did so because they were tougher, but that's too simplistic of a narrative. Foster "Tooter" Teague was one of the athletes who the papers of the day reported leaving camp due to injury. Teague became a TOPGUN fighter for the Navy, flying the F-8 and F-4 fighters during Vietnam. His résumé is littered with superlatives, including earning a Silver Star, commanding the aircraft carrier USS *Kitty Hawk*, and being selected as a pilot for a top-secret program to test out a Soviet MiG fighter. Gifted players like Teague, Broussard, and Boring didn't leave camp because they weren't able to handle it. Whether because of injury or priorities, mindless suffering in the dry heat lost its appeal. That's no more an indicator of their internal fortitude than an employee working long hours for minimum wage resigning to find a better opportunity.

And the players who lasted? They didn't do so out of some innate strength or resolve. Many did so because they had no other choice. Jack Pardee echoed a familiar refrain for those who made it through camp: "I never thought about quitting. . . . If I did, where would I go?" Running back Bobby Drake Keith summed it up best: "A lot has been made about the ones who stuck it out being stronger or whatever. But I think most of us survived because football was important to us for whatever reason, and it was in our nature to do whatever we had to do to stay on the team and stay in school. Our instinct was survival."

Success is complex. I'm not proclaiming that Bryant wasn't a great coach or teacher in many regards. But when it comes to developing toughness, we have to ask whether the Junction camp was successful or not. It accomplished precisely what Bryant wanted at the time: eliminating players after a coaching regime change. But did it develop toughness? The immediate performance gains, or lack thereof, suggest otherwise. And if it did work, then it did so for one-third of his team at best. It fundamentally failed the other two-thirds. Throw eggs at the wall. See which ones don't break.

We need to outgrow this old model. Even Bryant did. At a twenty-five-year reunion of the camp survivors, Bryant apologized to his former players, acknowledging that he had mistreated them. In his later years, he remarked that "if it had been me, I'd have quit a dozen times, but they never quit. I didn't know if I was doing it right or not, but it was the only way I knew how to do it."

Bryant's equating of handling extreme conditions with success has stuck around. Bryant and the story of the Junction Boys formed the basis of our model of toughness, setting the standard for how a generation defined it. It is, arguably, the origin story for this narrative: a Darwinian survival-of-the-fittest trope that is taking place in homes and on athletic fields across the country. Siphon off the weak; let the strong remain. Those who survive will thrive. Those who don't make the cut, well, they could find something easier to do. No water, go until you puke, harden players, develop thick skin. A sort of Machiavellian "the end justifies the means" concept.

How We Got the Toughness Story Wrong

Before Bryant was a football coach, he served in the Navy during World War II. The similarities between our common conception of military-style training and the Junction Boys style of toughness is striking. But it's also wrong. The great irony is that the military doesn't use boot camps and similar exercises to develop toughness. In actuality, the military is at the forefront in developing real toughness, just not in the way that most of us imagine. We just took the wrong lessons.

The Navy SEALs' "Hell Week" wasn't designed as a method to toughen up and develop soldiers; its goal was to separate those who could survive the rigor of war—to see who could handle the stress they were about to face; to see if, when in the foxhole, they could do their job. In sports, we took it as a means of development, and unfortunately, we made the same mistake off the playing field. We mistook the sorting portion of the military as development and looked right past how the military actually develops soldiers to survive extreme adversity. The old model of toughness, in essence, throws people into the deep end of the pool but forgets that we need to first teach people how to swim.

When we face extreme stress, we sometimes fall into a strange state, where our perceptions shift, memory fades, and we're incapable of acting. In this case, we're not talking about the stress of giving a presentation, but a more harrowing kind of stress. Think soldiers in the midst of battle, emergency workers experiencing catastrophe, or the trauma of physical abuse. Psychologists refer to this experience as dissociation.

Dissociation is the feeling of being detached, as if your mind has hit the eject button to get you through the experience. It can be separated into three categories: amnesia, depersonalization (feeling detached from self), and derealization (detaching from your surroundings). Our perceptions change; we forget, zone out, and feel incapable of action. It's an extreme involuntary coping strategy, a last-ditch effort for survival. And when you are in a life-or-death scenario, where the well-being of both yourself and those around you depends on you performing specific tasks, dissociation isn't the state you want to be in. Yet, that's the exact predicament our military faces: high stress and a need to perform.

We tend to think of experienced soldiers as strong, stoic, and tough. And rightfully so. Yet, according to research, when put through extreme stress, 96 percent of the soldiers experience dissociative symptoms. Sixty-five percent of experienced soldiers reported having "lost track of what was going on." All but two of the ninety-four soldiers interviewed said they "felt as if they were looking at the world through a fog." Not exactly the experience you want to be having when in the midst of combat and survival.

While nearly all soldiers suffer from the fog of warlike conditions, not everyone falls deep into the throes of dissociation. Some soldiers can stay engaged and maintain a calm, clear head during even the moments of utmost stress. They are able to keep their cognitive wits. The fog is still there, but they find a way to navigate through it. To soldiers, this capacity means the life or death of not only themselves but their entire squadron. The US military desperately needed to train this ability. But how?

During the first half of the twentieth century, survival training was relatively simple and straightforward. Learn what to do if your plane goes down, how to survive extreme environments and, if captured, hold out as best you can. It wasn't until after the mass casualties of POWs in the Korean War that survival training became strategic. In 1961, the Air Force opened their first survival school, with the Navy and Army following soon after. The SERE (Survival, Evasion, Resistance, and Escape) program was born.

There are three critical phases of survival training: classroom, evasion, and detention. The latter two are what receive most of the attention. Evasion involves getting dropped off in the wilderness with clear goals: evade actors dressed as enemies and survive off the land. Just as soldiers are getting used to foraging for food, they are captured, blindfolded, and whisked away for the final phase of their training. In a mostly classified POW experience, soldiers are locked away in cells and put through periodic physical and psychological beatings. One former Navy pilot reported having to endure a blaring speaker in the corner of the cell: "A mind-numbing cacophony of an out-of-control saxophone was followed by Rudyard Kipling reciting his poem 'Boots' over and over in a very haunting voice." Time spent in the cell is interrupted by interrogation sessions, which might include being stuffed into a box or another form of simulated torture, all while being told that your only way to safety and comfort is to disclose information.

Detention is the part of SERE training that is dreaded by those who have to go through it and romanticized by those who hear about it. By all accounts, the experience feels real. Your mind believes your life is in danger. And it largely works in developing resilience. But it's the often-overlooked first phase that distinguishes SERE training as a method of developing versus sorting.

Before being dropped into the wilderness, soldiers are trained. The initial classroom phase consists of a barrage of lectures designed to give soldiers the skills necessary to survive, evade, and resist. The US Air Force's SERE operations manual is 652 pages and covers everything from psychological aspects of survival to knowledge of medicine, methods of camouflage, fire making, and whatever in the world "river hydraulics" is. The psychology section includes how to handle boredom, loneliness, hopelessness, losing the will to survive, and over two dozen other maladies. In other words, the goal of SERE training is to prepare you for just about anything that you'd face, all before you even set foot in the woods or a simulated POW camp.

But aren't the harrowing conditions of the final two phases of training similar to a Junction Boys–style camp? The training isn't designed to push you to failure; it's to prepare you for a potential reality that you might face. SERE training is based on the concept of stress inoculation. If we "vaccinate" someone to extreme stress, they'll be able to handle it better. The first step isn't to throw someone into the deep end of extreme stress; it's teaching the skills necessary to cope with the situation. Without learning the skills, the second part—putting individuals in a harrowing environment to practice those skills—is useless. The key to stress inoculation, though, is like a real vaccine: you don't want it to be so powerful that you overwhelm the system.

SERE was just the start. In the latter half of the twentieth century, the military realized that making people do difficult things wasn't enough. Bear Bryant–style boot camps were great sorters but poor teachers. In 1989, the US Military Academy introduced the Center for Enhanced Performance, focusing on teaching cadets about goal setting, positive self-talk, and stress management. Soon after, programs in every branch of the military introduced mental skills coaching, culminating in nearly a dozen programs focused on improving mental strength and resilience. As of 2018, the US Army is the largest employer of sports psychologists in the country.

In 2014, the RAND Corporation was tasked with answering an important question, "Is the Air Force doing everything it can to prepare battlefield airmen to perform successfully under stressful conditions?" In evaluating nearly a dozen methods of preparing soldiers for the stress they'd experience, there were two items sitting atop the list of recommendations. First, emphasizing core skills that aid performance, including confidence, goal setting, attention control, arousal control, imagery, self-talk, compartmentalization, and mental skills foundation. And second, ensuring those skills are mastered before exposure to stressful conditions. In other words, you need to teach the skill first. Even the vaunted Navy SEALs recognized this distinction when in the early 2000s, they implemented a classroom phase designed to train candidates how "to monitor their psychological performance and learn to maximize mental toughness skills."

Research and practice are clear. Stress inoculation doesn't work unless you have acquired the skills to navigate the environment you will encounter. As sports psychologist Brian Zuleger told me, "Telling people to relax doesn't work unless you've taught people how to actually relax. The same goes for mental strength. The historical way to develop toughness was to do something physically challenging, and you'd have a fifty-fifty shot if they thrived. You have to teach the skill before it can be applied." Throwing people in the deep end doesn't work unless they've been taught the basics of how to swim.

Let's take the vaccine analogy a little further. What happens if you don't inoculate yourself against the specific stress you end up facing? If, like with the yearly flu vaccine, the doctors and scientists select the wrong influenza viruses and load up the vaccine with a strain that only offers partial immunity to this year's version of the flu, are you out of luck? Destined to get severely sick and possibly die? Of course not. You have a backup: your immune system. A healthy and robust immune system can fight many circulating viruses that it has little prior knowledge about or immunity to. Our immune system has both a general response to any foreign invader and an adaptive response of specialized cells tuned toward targeting pathogens. The adaptive response works better at ensuring safety from a specific virus, but the general response gives you a last line of defense in case of emergency. Within these two responses are a variety of methods to deal with whatever enters our body, even if we've never seen it before.

A tough individual is like a robust immune system. It's best to have knowledge to prepare for specific stressors, but even if we encounter an unfamiliar threat, we have a number of methods to cope with whatever comes our way. Recently in the military, there's been a push for not only utilizing stress inoculation, but also building a robust foundation for all soldiers. Comprehensive Soldier Fitness (CSF) is a program designed to "maximize [a soldier's] potential and face the physical and psychological challenges of sustained operations." *Sustained* is the key word there. If survival-style training is a vaccine, then CSF is maintaining a healthy and robust immune system. It focuses on developing resilience and skills to cope over the long haul.

Militaries worldwide have adopted a "strength-based" approach to psychological development. As noted previously, developing a "mental skills foundation" was one of the key skills in the review on the Air Force's stress inoculation program. Pulling from the field of positive psychology instead of focusing on preparing for the worst and shoring up weaknesses, such an approach teaches the basics of well-being and mental health, including learned optimism, resilience, post-traumatic growth, and emotion regulation. The goal is to develop skills that help handle the specific stresses on the battlefield and the stressors that hit us in everyday life. In 2015, the US Army enacted the Human Dimension Strategy, aimed at developing the holistic person, instead of the old model, which emphasized technical and tactical knowledge. Among the objectives for the program: intellectual optimization, social intelligence, holistic health and fitness, decision making, and team building.

We took the wrong message from the military on toughness. No, it's not that we need to put athletes through ridiculous training or enact authoritarian control. It's not the discipline or demandingness we needed to copy. Not even the strength or machismo. We took the sorting to mean training. We saw the training but forgot the teaching. We glossed over that the training wasn't hard for the sake of creating toughness. It was designed to simulate and train for the actual demands soldiers would face on the battlefield. The lesson wasn't that we just need to put people in difficult spots and force them to deal with adversity. We need to teach them how to navigate the discomfort they'll soon face. We took the lessons from the 1940s military, ignoring that even the military doesn't subscribe to the Bear Bryant model of toughness anymore.

Toughness isn't a sorting exercise. We need to teach the skills to handle adversity. Development is not merely putting people through challenging times. As the military discovered, sink or swim doesn't work. When researchers evaluated soldiers who were able to keep a clear head during extreme stress, they found that the soldiers:

- appraised stress as a challenge instead of a threat, thanks in large part to a better assessment of what they encountered;
- utilized a diverse array of methods to cope with stress, demonstrating a high degree of cognitive flexibility;

- processed internal signals better, without reacting to them; and
- didn't react to negative stimuli but instead were able to change their physiological state.

In other words, soldiers were training their biology and psychology to work in tandem during challenging moments. It's not that they weren't experiencing discomfort; instead, they had figured out ways to maintain clarity when everything around them was pushing them toward chaos. High performers are able to work their way through adversity and challenge with the same equanimity. When put in situations that require toughness, it's not that they are bulldozing through the experiences; they navigate them with grit and grace.

The best of the best have another factor in common. No, it's not that they were born with extraordinary abilities to manipulate their inner world to handle adversity. It's not that they are immune to stress and anxiety so that they can work their way through situations easier than you or I. Their secret? When dealing with discomfort, they all want to quit. Under extreme levels of discomfort, our biology and psychology push even the toughest of us to give up. I surveyed dozens of authors, entrepreneurs, executives, soldiers, and athletes, nearly all having moments of wanting to throw their manuscript in the trash, debating ways to get out of their approaching deadline, or finding a hole to step in to end the misery of the race they are running. Negative thoughts of quitting are normal. They don't mean you are weak. They represent your mind trying to protect you.

We all face an inner battle, a slew of feelings, emotions, and thoughts that push and pull us toward persevering through and throwing in the towel. Sometimes our inner world screams at us to quit; sometimes it nudges us toward apathy and complacency. Jonathan Wai, a prolific academic at the University of Arkansas who studies gifted education, described to me how he often loses this inner battle: "I find myself staring off into space, avoiding writing and revising various papers and projects. . . . I often side with the lower-hanging fruit." We all face similar pulls, and understanding this process is central to navigating it.

Choosing the Difficult Path

It's Friday night. You're sitting in your apartment, waiting patiently. You have plans, or at least you think you do. Earlier in the day, your new love interest said he'd text early that evening to set up a date. Five o'clock, six, seven, all come and go without the familiar buzz of your phone indicating some sort of notification.

A slight feeling of unease begins to creep through your body. Your shoulders tense as anxiety begins to build. These initial sensations act like a small blinking light on your car's dashboard, sending the signal to your running mind, "Hey, something is going on. Maybe I should pay attention to it." You brush it off. "He's probably running late from work. He'll text soon." Meanwhile, glances toward your phone increase; once every fifteen minutes turns into every five minutes and finally becomes incessant checking. Early on, the TV provides a worthy distraction, but as the night wears on, your attention gravitates to that rectangular piece of technology that holds the key to modern communication. As unease turns into anxiety, your world narrows. Distraction fails. Your mind tunes to only one thing. The alarm inside your head shifts from blinking to blaring, from warning to impending disaster.

A calm inner dialogue transforms into a raging internal debate. An angel and devil emerge, pushing and pulling us toward one action or another. "Maybe I should text him? No, give him space. Maybe Snapchat him instead? That would be too needy. He said he would text. Just wait." You feel like a lunatic, spiraling out of control, in an illogical fight against yourself. Your imagination runs wild, and visions of him standing you up, or maybe even ditching you for another date, surface. You've spiraled into a full-blown freak-out.

At this point, your inner turmoil moves toward actual behavior. Barrages of texts or calls to your lost love interest soon follow. We've all been there, men and women alike. It started innocently enough with an expectation that he or she would text. From there, it spiraled down, down, down, until we've become an irrational mess. We've transformed from respected adult professional to the despairing depths of our teenage years. We are a wreck.

Whether you're trying to resist the pull to quit or barraging your potential lover with texts, the key to toughness lies in navigating this biological and psychological cacophony. I've hinted at this sequence throughout this chapter, but let's clearly define it. How do we get from discomfort to action?

Feel → Inner debate → Urge → Decision (freak out OR find our way through)

We experience some sort of sensation or feeling while our mind runs rampant with internal thoughts that push and pull us in different directions. Both our feelings and thoughts nudge or urge us toward some sort of decision. To quit, persist, or change our goals before we ultimately make a decision.

Instead of seeing these as distinct steps, see them as a mishmash that work together. Sometimes we can distinguish each experience: feeling sad, then thinking about it, then having an urge to act because of it. Other times, we jump straight from the slightest sensation to a decision. But each segment affects the rest. A touch of fear can spiral out of control if accompanied by a rash of "what if?" thoughts. When we're exhausted, or experiencing high levels of stress and anxiety, we're more likely to take the quick route, finding an exit in whatever way possible. We choose the easiest path.

According to the latest scientific theories, the brain functions to maintain order. Our brain is an uncertainty-reducing machine, willing to do whatever it takes to minimize surprise, even if it is at a high cost. Whenever we face something that throws our internal state off-kilter, we go about trying to solve it. We seek out a solution that moves disorder to order. Sometimes that means giving up, such as when we are a third of the way through a project and can't see the finish on the horizon. We often quit so that the unknown becomes the known. Other times it means changing our expectations before even beginning a task. Or it could mean exploring, accepting, or avoiding whatever it is that has led to unease or discomfort. Uncertainty demands a conclusion. We have an innate need for closure, however we can reach it.

Toughness is about making the pull for closure amid uncertainty work with you, not against you. It's training the mind to handle uncertainty long enough so that you can nudge or guide your response in the right direction. To create space so that you don't jump straight from unease to the quickest possible solution, but to the "correct" one. The first step in redefining toughness is to understand where we went wrong, why bulldozing through often leads to a worse outcome. In the rest of this book, we will unpack the pillars that allow us to address every part of this cycle from what we feel all the way to the decision we make.

THE FIRST PILLAR OF TOUGHNESS

DITCH THE FACADE, EMBRACE REALITY

CHAPTER 3

Accept What You Are Capable Of

In 1966 President Lyndon B. Johnson created the Presidential Physical Fitness Test. It consisted of a series of physical challenges determined to assess one's athletic ability and to encourage grade school children to be physically fit. For much of its history, the mile run was a staple of the test.

When I ventured down to Haude Elementary School to watch a group of students tackle eight laps of a dirt 200-meter track, it was unlike any of the thousands of college and professional races I've watched over the years. When Mrs. Passmore yelled, "Go!" the kids bolted off the line in a dead sprint for the lead. As they made their way past the first lap, reality hit: they had a long way to go. Kids slowed dramatically, transforming from a sprint to a jog. For those in the middle or back of the pack, their transition was even more abrupt, from a run to a walk. While a handful of the kids held on to a consistent slow trot during the middle laps, the rest of the field found themselves alternating between walking and short but quick bursts of running, often triggered by the teacher or a friend cheering them on. Inevitably, as the finish line neared, they all summoned up a final kick, a sprint that rivaled their initial burst of speed off the line. The best way to describe the race was like a yo-yo bouncing back and forth between running and walking. Unlike their younger counterparts, experienced endurance athletes adopt a more optimal, even pacing strategy, staying nice and steady. How do we make that transition, abandoning our sprint/walk strategy in favor of the preferred, more even-paced approach?

When it comes to just about any endurance task, regardless of whether it's cycling, swimming, or running, we utilize a simple metric to fine-tune our pacing: the sensation of effort. Current theories suggest that we have a kind of internal map, approximating how hard a race should feel at any given moment. Our mind knows that if we are attempting a longer race, like a marathon, the first few miles should feel relatively comfortable. If they feel harder than expected, it's a message telling us to slow down or else we might be in trouble later. We utilize effort as a kind of internal gas gauge, comparing how quickly our car is burning through fuel versus how far we have left to drive. In the field of exercise science, there's a simple formula that dictates how we utilize effort to govern our pacing and ultimately our performance:

$$\text{Performance} = \text{Actual demands} \div \text{Expected demands}$$

If the pace feels much more comfortable than you expected, you pick it up. If it feels more challenging than expected, our feeling of pain and fatigue will go up, our inner dialogue will become negative, and we're more likely to slow down. If you reach mile 4 of the marathon and you're already out of breath, you might spiral to a full-blown freak-out and drop out. At each point along the way, our brain makes an internal calculation of whether this feels harder or easier than expected. Pain and fatigue are our body's way of nudging us toward a course correction. We can't sustain the pace, so we'd better slow down. If we don't listen, our body will take matters into its own hands, shutting us down to prevent catastrophic failure or damage—like a car that may run out of gas miles before its destination.

A tougher runner isn't one who is blind with ambition or confidence, but one who can accurately assess the demands and the situation. The magic is in aligning actual and expected demands. When our assessment of our capabilities is out of sync with the demands, we get the schoolchildren version of performance. Starting a project with reckless confidence, only to look up and realize the work it entails. When such a mismatch exists, we're more likely to spiral toward doubts and insecurities, and to ultimately abandon our pursuit. When actual and expected demands align, we're able to pace to perfection, or outside of the athletic realm, perform up to our current capabilities. It's why experienced writers don't go into their first draft expecting perfection. They understand it's going to be messy, and often not that good. Contrary to old-school toughness wisdom, a touch of realistic doubt keeps us on track and makes it more likely that we will persist.

Toughness is about embracing the reality of where we are and what we have to do. Not deluding ourselves, filling ourselves with a false confidence, or living in denial. All of that simply sends us sprinting off the line, only to slow to a walk once reality hits. Being tough begins long before we enter the arena or walk on stage. It starts with our expectations.

The Threat of Death

Climbing a ladder to reach the roof of your home isn't the most dangerous endeavor. A well-made ladder is sturdy and durable, equipped with steps to make the short climb to the top as uneventful and steady as possible. Despite the relative safety of such a trek, the first time we get beyond the bottom few steps, a wave of unease rushes through our body. Maybe even a tinge of fear and thoughts of "What if I fall?" pop into our mind. If we look down as we near the top steps, these feelings magnify. We might rationalize our experience, believing that a fall is unlikely. And even if we did misstep and plummet to the ground, we'd likely be bruised but survive. As we come to terms with our role of handyman and adjust to climbing ladders, the fears and feelings dissipate. The anxiety and fear slowly fade. We feel safe.

Now, imagine instead of looking down from ten feet and standing on a secure metal step, you peer down to the ground that you can barely see, thousands of feet below. Instead of a secure step, you're on a rock face with nearly zero outcroppings or ledges. Feet and arms perched on minor deviations of the rock surface, holding on to parts of the rock that may jut out from the surface by a few millimeters at best. You aren't grasping on to a ledge with the entirety of your hands; it's the very tips of your fingers that secure you to the granite wall's surface. And unlike any rational human being when attempting to climb a vertical wall, there is no safety net. No ropes to anchor you to the side of the wall, to protect you from falling to your death in the case of a misplaced finger or foot. Just you, alone on the wall. Welcome to the world of free soloing.

In June 2017, climber Alex Honnold faced such a challenge, climbing the face of El Capitan, a three-thousand-foot vertical granite monolith in Yosemite National Park. He did so without a rope or safety harness, with the narrow grip of his hands and feet keeping him from falling to his death. To tackle such an endeavor, mastery of climbing is a given, but how does someone deal with the fear, anxiety, and pressure of such an undertaking? When most of us stare down from the third-story balcony of our hotel room and experience fear from leaning over the edge, how is

Honnold able to master his emotions and inner dialogue to handle the challenge? His nickname, "No Big Deal," provides a clue.

Neuroscientist Jane Joseph took a peek inside Honnold's brain to see if she could find an answer. While lying in an fMRI machine to scan blood flow in his brain, Honnold watched a series of disturbing images flash in front of his eyes. Think disfigured bloody corpses or a toilet filled with feces. Pictures designed to make just about anyone cringe. Even if we have no visceral experience, for even the strongest among us, our brain will betray us with an internal sign of provocation. An almond-shaped part of our brain called the amygdala should light up. The amygdala has many functions; primary among them is to detect and respond to threats. When disgusting or threatening pictures—like those shown to Honnold—trigger the amygdala, a cascade of events is unleashed that eventually result in a slew of hormones released and nervous system activity. We call this a stress response.

In a conversation captured in the magazine *Nautilus*, Honnold asked whether or not the images of children burning counted as stress. Despite being reassured by Joseph that such images routinely elicit some sort of emotional arousal, even in rock climbers and adrenaline junkies, Honnold quipped, "Because, I can't say for sure, but I was like, *whatever*." And as Joseph would later see, Honnold wasn't putting on an act. His brain echoed his experience. There were no flashes of color to indicate activity in the brain's threat- and fear-sensing areas, just gray. Honnold's amygdala didn't react to a single disturbing image. Not a blip of activity. Honnold's secret weapon might be that his emotional reactivity is monk-like. When the rest of us are smashing the panic button, heading toward a freak-out, Honnold's mind is enjoying the scenery, quietly thinking, "There's no threat here."

Honnold isn't superhuman. Shortly into his first attempt to climb

El Capitan, Honnold mused, "This sucks. I don't want to be here. I'm over it." He pulled the plug, explaining, "I don't know if I can try with everybody watching. It's too scary." It's not that Honnold never experiences threats, that his amygdala never lights up. It does so when he needs it to. That day, fear rang out, and he listened, pulling the plug before disaster struck. He'd wait to reach his goal another day.

Through a bit of luck, the right genes, and countless hours of mental and physical rehearsal, Honnold has fine-tuned his threat-detecting machinery to be triggered when something is truly off. Not when pictures pop up on a computer screen, but when he can't actually complete the task he sets out to. Our body's alarm system is malleable. We don't have to be monk-like and turn the knobs to adjust our sensitivity. We just have to get better at predicting.

Research consistently shows that tougher individuals are able to perceive stressful situations as challenges instead of threats. A challenge is something that's difficult, but manageable. On the other hand, a threat is something we're just trying to survive, to get through. This difference in appraisals isn't because of an unshakable confidence or because tougher individuals downplay the difficulty. Rather, those who can see situations as a challenge developed the ability to quickly and accurately assess the situation and their ability to cope with it. An honest appraisal is all about giving your mind better data to predict with. Like an epidemiologist predicting the public's response to a novel virus, a better appraisal allows us to unleash the response we need for that situation and moment.

Appraising the World Around Us

Whenever we face a new or stressful situation, our body does its best to prepare for what's to come. But we don't wait to see exactly what that noise in the brush is, or whether or not our job is dependent on the boardroom presentation. Our body cheats. Instead of waiting to see whether a task is hazardous, our brain makes its best guess about what we need to survive or thrive. It's the reason you feel nerves or a racing heart well before you step onto a stage. It's why, while waiting for the plane to reach the right altitude to jump out, novice skydivers are filled with dread, while veteran skydivers are excited. Inside their bodies, the novices are secreting cortisol, while the veterans have more adrenaline. Same event, yet the body releases different hormones to prepare for what's to come. Whether it's climbing a ladder or a mountain, our biological response and the sensations that come with it are guided not only by the actual experience but by our expectations. How we see the world shapes how we respond to it.

Before we step on stage or up to the batter's box, our body has already entered a performance state. The "rush of adrenaline" we feel is to prepare us for jumping out of a plane or stepping up to the starting line. And as that saying hints at, the experience is caused by the internal milieu of nervous system activation and hormonal release. Our body uses these chemical reactions to prepare for whatever we're going to face.

A stress response prepares us for action. We tend to think of it as a binary choice: fight or flight. But the reality is we have a slew of different ways to prepare for whatever is in front of us. It's the combination of hormones and nervous system activity that pushes us toward responding in a certain way. Adrenaline surges to prepare us for rapid movement. Simultaneously, the hormone oxytocin pushes us toward a group effort to get through the danger, and cortisol liberates vital fuel sources in our cells to prepare our muscles and minds to work over the next few hours. With each shift in the amount or type of ingredient in the recipe, we change everything from the texture to the flavor.

Some responses prepare the muscles for actions, others open up or slow down the blood flow, while some marshal the immune system to prepare for potential damage or injury. It's as if our body has its own 911 system, deciding whether to send an ambulance, police officers, firefighters, a social worker, or the SWAT team. How does it know which one to send? It's our evaluation of the situation and ourself that largely determines which way we'll go: protect or attack it head-on.

Are we like the expert marathoners who are able to match expected effort at any given moment during a race with their actual experience? Or are we like the grade school kids who misjudge how difficult running a mile will be? Does our appraisal of our skills match our appraisal of the demands of the situation? This mismatch between situational demands and our capacity to cope doesn't just determine running performance; it determines what kind of response to stress we'll have. When we see a stressful situation as something that could cause physical or psychological harm, we're more likely to experience a threat response—a rush of cortisol and a shift toward defending and protecting. Our body unleashes a threat response when we face a demand that we aren't quite equipped to handle. Like our novice skydivers, we're just trying to survive. How do we make it out of this situation with our physical and psychological health intact? We take fewer risks, playing to not lose instead of playing to win.

On the other hand, if we see the stressor as an opportunity for growth or gain, as something that is difficult but that we can handle, we're more likely to experience a challenge response. Instead of relying mostly on cortisol, our body releases more testosterone and adrenaline. We shift toward figuring out how to win the game, how to accomplish our goal. It's not that challenge or threat responses are good or bad; they each have a specific purpose. If we encounter a bear protecting its cubs on a trail, we want a threat response, freezing, assessing the situation, and then calmly backing away. But when we're trying to perform at our best, seeing the world through the lens of a threat isn't what we need. We want to see the task as a challenge. How we appraise ourself and the situation shifts which way we go.

If expectations partially determine what we feel, think, and do, is it best to downplay the difficulty? To tell ourself that the task won't be that hard or painful? When our expected effort is different from the actual effort, our brain course-corrects. The degree of mismatch determines the reaction. If there's a large mismatch between expectations and reality, our brain overcorrects. If we went into a test thinking it would be a cakewalk, at the first sign of difficulty, it's as if the brain goes, "Hey! What is this? This isn't supposed to be hard!" As a result, instead of figuring a way through the discomfort, we shut down. We go into full protective mode. We freeze.

Is it better to prepare for the worst? Go in with the anticipation that this race, presentation, or project will be the most difficult and demanding task that we've ever faced? In our racing scenario, if we expect an immense challenge and it's easier than expected, we'll pick up the pace and perform better! Wrong. If our expectations swing too far in the other direction, our brain goes into what I call "What's the point?" mode. The task will be so far outside of our capabilities that there's no point in using our full reserve to take on the challenge. We're doomed before we start. Our ability to be "tough" and handle adversity starts well before we even encounter any difficulty. It starts with embracing the reality of the situation and what you're capable of.

TOUGHNESS MAXIM

Our appraisal of a situation as a threat or as a challenge depends on the perceived demands of that stressor versus our perceived abilities to handle them. Do we have the resources to handle the demands?

Facing Reality

"It's easy to be tough when you know you can handle the situation. The true test comes when you can't," a former athlete of mine, Drevan Anderson-Kaapa, relayed to me about his experiences. Drevan was a three-time conference champion in college who also represented his country at the World University Games. But Drevan didn't just win races; he figured out ways to get the most out of his body, regardless of the situation. On the University of Houston track team, he became a sort of living legend, with stories about his heroics being passed down to every incoming class of students. There was the time after he'd won his individual race at a championship that he then volunteered to run the 4x400, an event he had not run in his entire collegiate career. Not only was it unfamiliar, but with team scores being a near dead heat, whoever won the relay captured the team title. Despite the pressure of having to win, he walked up to two of the coaches, who happened to be track-and-field legends Carl Lewis and Leroy Burrell, and declared, "I'll run it, and I'll run anchor," putting himself up against some of the fastest runners in college track. Of course, he sealed the victory, roaring back in the last 50 meters to claim the team title.

But it wasn't just his track accolades that set Drevan apart. During his college career, he not only competed at an elite level, but he also obtained a master's degree and was part of ROTC. Upon graduation, he went from representing his country on the track to serving it in the military. When it comes to toughness, there are very few people I've encountered who epitomize the word. Drevan is one of them.

As we sat at my kitchen table in the fall of 2018 and discussed toughness, he continued with his observations from the athletic and military world: "Everyone wears a mask. We carry around a facade, projecting an outer image of who we want to be. But when you are under stress, that fades away and you're left with what's underneath. Stress exposes you." When I asked Drevan to explain what he meant, he outlined two distinct masks he'd observed during his athletic and military days. First, there was the individual who looked the part, projecting bravado and self-assurance. They were the athlete who talked big, while downplaying the difficulty of the task at hand. "This will be a piece of cake," they'd say pregame. Yet, the moment something went awry, all that self-assurance faded away. They became timid, unsure of their abilities, and would fall back when challenged during the most difficult part of the race. The second version initially appears similar. They, too, have a sense of confidence, exuding a certainty that they'll be able to handle what's thrown at them. But standing on the starting line, they weren't downplaying what was to come, but were clear-eyed on the difficulty of the task at hand.

One example stood out. When dropped off in the middle of the woods as part of survival training, with minimal supplies and working off extreme sleep deprivation, Drevan and his companions were forced to put their skills to the test. As energy levels dissipate, your world narrows to focus on satisfying your basic needs: food, water, and sleep. Not much else matters. It's easy to forget the group and focus on yourself. Drevan found himself watching a number of his companions who days earlier were team-focused and extremely confident of their abilities succumb to the stress of survival. These individuals took more than their share of limited rations or snuck a few extra minutes of sleep when they were supposed to relieve their colleague of watch duty. The rest of the group maintained their discipline, keeping their cognitive wits and team focus, despite the situation. Under normal conditions, toughness is easy. Under extreme duress, not so much. We default to choosing the easy path.

When I asked Drevan what the difference was in individuals who were able to keep their cool, he replied, "When there's a difference between what you project and what you are capable of, it all crumbles under stressful situations. If, on the other hand, you're honest with yourself, and acknowledge what your strengths and weaknesses are, what you're capable of and what might scare you, then you can come to terms with what you're facing and deal with it. The you walking the streets and the you stranded in a jungle aren't that much different. So you're able to assess the situation with clear eyes and expectations instead of trying to live up to some false standard. It was the same thing in track. The ones who thought they had to bring 110 percent to a race inevitably fell apart. The ones who lined up on the starting line and thought 'This is going to be easy' or 'This is going to be extremely hard' didn't perform up to standard. They were living in an altered reality. The ones who said, 'Here's what I'm capable of. Here's what the race demands. I'm going to execute based on those two things.' Those are the ones who consistently perform."

Drevan's experience isn't unique; it's backed by scientific research. At the turn of the twenty-first century, a group of scientists analyzed soldiers across the military branches as they went through the same type of survival school that Drevan experienced. They noticed the same phenomenon. While nearly everyone experienced high levels of stress, one group seemed to zone out and become almost detached from the experience, while another was able to largely maintain a clear, level head and perform up to their potential. Both groups were managing stress in a different way, yet the ones who focused on the reality of the situation were better at handling various elements of the mission. These more robust soldiers appraised the upcoming task as a challenge instead of a threat. In concluding their work, the group of researchers summarized their findings that the more robust "individuals are more accurate in descriptions of what they encountered during stress." They were facing reality.

TOUGHNESS MAXIM

Embrace reality. Accurate appraisal of demands + accurate appraisal of our abilities.

Setting Your Mind on Reality

A key component of real toughness is acknowledging when something is hard, not pretending it isn't. An honest appraisal of ourself and the situation allows us to have a productive response to stress. It can shift whether our body is pushed toward fear or excitement, challenge or threat. And in turn, whether we'll take a risk, shy away, or be able to access our full potential. Embracing reality isn't about sitting around and thinking about what you are capable of and whatever challenges you are about to take on. There are research-backed strategies that we can use to fine-tune our mind, to nudge us toward an accurate appraisal, and even more so toward a productive stress response that prepares us for action.

1. Set Appropriate Goals
We're often told to shoot for the moon or dream big when it comes to setting goals, but research points to the opposite conclusion: set goals that are just beyond our current capabilities. If there's too large of a mismatch between our abilities and our goal, our motivation decreases. It's as if our brain shrugs its shoulders and says, "What's the point? We aren't going to win anyway." Whenever we set our expectations too high, we're more likely to enter the "freakout" stage of our toughness pattern. Instead of going big, set just manageable challenges.

2. Set Authentic Goals
In a series of studies in the Netherlands, psychologists sought to understand why some individuals are able to make progress and reach their goals, while others continually fall short. Over three studies, researchers found that better goal authenticity contributed to better goal achievement. When people chose goals that reflected their true selves, not their public selves, they were more likely to follow through. Those who failed often chose goals that were imposed on them by a parent, coach, or society in general. For those

who were successful, goals came from within, reflecting who they were and what they cared about. A high degree of self-knowledge is what allowed these individuals to see clearly.

Seeing reality doesn't just mean understanding what you're capable of and what the actual task demands are. It means taking the time to understand who you are and what matters to you. Whether through introspection, journaling, or conversations with close friends and family, do the hard work to ask what matters and why it does. Tough people are self-aware, and they get there by embracing reality and understanding who they are.

3. Define Judgments and Expectations

If we all took the viewpoint of Ricky Bobby from the movie *Talladega Nights* that "If you aren't first, you're last," we would be in trouble. What happens when we set an audacious goal and quickly realize we can't hit it? When we don't think we have a shot, whether it's in a race or in class, our brain shuts us down. Our mind jumps into protection mode, thinking, "Well, we aren't going to win, so why waste any energy trying?" We are unintentionally killing our drive if we define success and failure in such a narrow way.

I've witnessed far too many people harm their motivation and their performance by defining success in the wrong way. They'll focus solely on outcomes, neglecting to realize that what place they finish in or what their grade is on a presentation is largely out of their control. Shifting the focus toward process-oriented goals, such as the effort you can put forth, helps remedy these situations. It also provides vital feedback that allows you to grow in the future. When you judge yourself solely by what place you came across the finish line, it provides zero actionable information on how to improve in the future. Judging yourself by how much effort you gave or whether you executed your plan offers a road map for what can be worked on during the next go-round.

4. Course-Correct for Stress

A group of French researchers from the University of Nantes wanted to see how stress affected individuals' judgment of what they were capable of. They chose a simple task: estimating how high of a bar one could step over. The trick was that participants had to make their guess after being kept awake in a lab for twenty-four hours straight. Sleep deprivation does a number on the brain, inducing stress and fatigue. Regardless of the actual height they could navigate in a normal state, participants severely underestimated the height they could step over when in a sleep-deprived state. Stress alters our judgment of what we're capable of. In another study, researchers found that those in chronic pain tend to overestimate the distance to walk to a target. These findings led sports psychologist Thibault Deschamps to state, "Individuals perceive the environment in terms of the costs of acting within it."

In another study, researchers asked individuals standing at the bottom of a hill to guess how steep the hill was. The overwhelming majority greatly overestimated the steepness, guessing it was a 20-, 25-, or even 30-degree-grade hill. In reality, the hill had a 5-percent grade. There was one group of students who were accurate: the cross-country team. If we have the capability to run up the hill without too much undue stress, we see the steepness for what it is. If we don't, it looks overwhelming. In an interesting twist, when researchers took those runners and made them go out for a long run and then come back and judge the slant of the hill, their hill-judging expertise disappeared. In a state of fatigue, they started overestimating the slope to a much greater degree.

Fatigue had shifted their capabilities and, with them, their perception abilities.

When we flip into a threat state, a freeze reaction, or a full-blown freak-out, the normal often seems unattainable. We tend to overcompensate, greatly diminishing what we're capable of. Part of having an accurate appraisal is course-correcting. If you feel tired, fatigued, or anxious, you can learn how to navigate that. But knowing that you might sell yourself short gives you the power to do something about it, and to readjust as stress or fatigue mounts.

5. Prime Your Mind

In 2018, a group of researchers out of University College London wanted to see how stress impacts the way we treat information. In the study they looked at how on-duty firefighters and students who were about to step on stage to give a speech handled being informed of either good or bad news. For example, being told your chance of getting in a car crash or suffering serious injury in a fire was much higher than the subjects thought. When relaxed, participants tended to ignore the bad news and embrace the good. Hearing that there was a greater chance to suffer some negative consequence didn't impact their behavior or mood. But under stress, as lead researcher Tali Sharot summarized, "They became hyper-vigilant to any bad news we gave them, even when it had nothing to do with their job (such as learning that the likelihood of card fraud was higher than they'd thought), and altered their beliefs in response."

Stress shifts us toward a negative bias, priming us to search out and recognize danger or threats in the environment. This is a great evolutionary survival mechanism, but it can hamper our performance when we aren't really in danger. To combat this quirk of evolution, prime your mind to search for opportunities, not threats. In *Peak Performance*, I outlined research that shows that when athletes warm up by "doing what they like," they alter their hormonal state in a positive manner. The same phenomenon applies to artists and executives. The closer you are to a performance, the more you want to prime with what you're good at. Reviewing mistakes, working on weaknesses, telling yourself that you "can't hit the slider, so watch for it" backfires when you're on deck. Those are items you work on far before it's time to step into the batter's box.

***

True toughness begins long before we reach the playing field or the conference table. It starts with how we assess the situation and our capabilities. Our expectations set the stage for our biological reaction. Our assessment can affect every step along the toughness chain, biasing us to feel more pain and leading to a premature freak-out. Before we get to the point where we need to bounce back off the turf after a brutal tackle, we need the right frame of mind. Our biology is already primed and biasing us in a particular direction before facing any sort of challenge. What direction we end up going is largely up to how we see ourselves and the world around us.

CHAPTER 4

True Confidence Is Quiet; Insecurity Is Loud

Leonard "Buddy" Edelen was the consummate professional. As an up-and-coming long-distance runner in search of Olympic glory, he was meticulous to the extreme. Every morning, he recorded his heart rate upon waking and made a note of how many hours he had slept. He tracked his weight, his workouts, and how he felt after completing them—all in the pursuit of optimizing his performance and making sure he was staying on the right side of the fitness-fatigue balance. He'd send off this trove of data and notes to his coach, Fred Wilt (a man who worked full-time as an FBI agent even though his true passion was distance running), who would offer comments and suggestions on where he'd gone wrong and how he could improve.

This unique coaching partnership led to astonishing accomplishments. Edelen's breakthrough came when he knocked almost a minute off the world record in the marathon, becoming the first American to hold the record in nearly four decades. It was a triumph of the painstaking and professional approach that Wilt and Edelen cultivated.

You might be asking why you don't know who Buddy Edelen is. Wilt and Edelen didn't correspond via email; they did so via postage mail. Edelen broke the world record in 1963, a few years before the running boom ushered in the craze of jogging. Despite the professional approach to running, Edelen was a full-time schoolteacher, completing much of his training as he ran to and from work every day. Despite his lack of fame and running's relative lack of popularity, Wilt and Edelen had done the unthinkable for the time: turned an American into the best runner on the planet.

Wilt was on the cutting edge of the physiology and psychology of training. In a time when training knowledge was in its infancy, when fear of training too much prevailed, Wilt was an innovator. He wrote to the best runners and coaches in the world, collecting samples of training from across the globe. He procured a hypnotist to change his athlete's mindset to see pain as something to embrace instead of avoid. Wilt and Edelen pushed boundaries. Yet, despite their success, human nature still reared its head.

In one of their mail correspondences, scribbled in the margins next to an entry detailing a run two days before an important race, Fred Wilt wrote, "I can't say this 40-minute jog will hurt you. I can say it does not help two days before a race. This is a manifestation of uncertainty. There is a time to train and a time to rest—not halfway rest. This is a bitter lesson you have not accepted." A coach scolding his prized pupil. Wilt recognized that despite reaching great heights, Edelen had a glaring insecurity. His obsessiveness—manifested in high training volumes and tracking everything imaginable before the quantified self movement had made it trendy—was also holding him back. He didn't have the confidence to step back, to rest.

Edelen told *Sports Illustrated* in 1964, "No one really accepts you as being sane if you run as much as I do a week. But if I rest a day or two after doing this tremendous amount of exercise, I feel very irritable and nervous. It's as if something has been stolen from me. Training gives me a feeling of tranquility." Edelen needed to run, not just to improve his fitness but to quell the doubts.

Athletes fear that if they are not training, they are falling behind, their fitness slowly seeping out of them. This phenomenon is quite common in all types of high performers. The entrepreneur whose insecurity leads him to incessant "grinding," the CEO who can't step away on weekends for fear of falling behind, or the creative who puts off publishing their work until it is just right. We often mask our insecurities with perfectionism and extreme levels of work. Buddy Edelen, a man who cemented himself in the annals of history by running faster than any man in the history of the world at the time, a man who won the 1964 Olympic Marathon Trials by nearly twenty minutes in scorching ninety-degree heat during a time when drinking water was frowned upon, lacked the confidence to follow his coach's plan and rest going into a race. If a man such as Edelen, who exuded tenacity, lacked confidence and couldn't control his insecurity before a race, how in the world can we mere mortals develop that capacity?

Doubts and insecurities are part of being human. Even if you're the best in the world. We all struggle with but want confidence, that sense of assurance that we'll be able to prevail at whatever we do. When we lack confidence or belief, our insecurities and doubts have room to grow. They move from subtle reminders that we need to align our expectations with reality to a constant reminder that we aren't fast, strong, or smart enough. As doubts take over, our perception of our resources shrinks and a mismatch between what we think we're capable of and the task before us emerges. We end up going on short jogs to quell the doubts, instead of doing the wise thing and resting. Confidence plays a crucial role in toughness, acting as the counterbalance to our natural insecurities. Confidence keeps our doubts in check, freeing us up to perform to our full capacity. Yet, the ways in which we've traditionally tried to instill belief largely do anything but.

Confidence appears simple: believe in yourself. A phrase that is plastered in classrooms throughout the country, and something that just about every parent, coach, or teacher has uttered. The old model of confidence focuses on the outside. Crafting the appearance of someone who looks like a strong, self-assured individual. We tell our children to believe in themselves, without explaining how to develop that belief. We've fallen for the Instagram version of confidence, emphasizing the projection of belief, instead of working on the substance underneath. We need a new approach to building confidence, one focused on the inside.

Shaping How You See the World

In 2009, performance psychologists Kate Hays and Mark Bawden had the opportunity to sit down with fourteen of the most accomplished athletes in their sport and ask them about their highest highs and lowest lows. Thirteen of the sample had won medals at a major championship (e.g., Olympic Games), and the lone individual who hadn't was a world record holder. This unique glimpse into the mind of the best wasn't just an interview; it was part of a study for the English Institute of Sport examining the role of confidence in those who had reached the pinnacle of their sport.

We tend to think of the best of the best as impervious to the feelings of doubt and insecurity that you or I might experience. As I've worked with world-class performers across a variety of domains, I've noticed one consistent theme: they are human, just like the rest of us. They aren't emotionless machines immune to the effects of pressure or poor performance. And that's precisely what Hays and Bawden found in their research. Despite achieving at the highest level possible, all could point to debilitative periods where their confidence waned and their performance suffered because of it.

It's not just that high performers suffer lapses in confidence but that it infiltrates and affects their thinking, feeling, and actions. When confidence was low, the athletes "were irrational, and unable to control their nerves, think positively or maintain focus on their usual routines." It's as if their brains were hijacked. Their worldview turned dark and gloomy, and simple tasks became difficult. Or as Dr. Hays and Dr. Bawden found, athletes suffered from a triumvirate of symptoms: faulty cognition, negative affect, and ineffective behaviors. They could not maintain focus as their attention drifted toward what others were doing or got hijacked by the doubts taking over their minds. They experienced a wider range of negative emotions, including nervousness, unhappiness, and an inability to enjoy the competition. Joy and thrill turned into anxiety and despair. They began to see the competition as a sign of a threat, not a challenge. And most importantly, when confidence was low, their behavioral responses followed their cognition and emotions. They were timid, indecisive, and withdrawn, and they lacked that extra bit of fight they normally possessed. Despite being some of the most accomplished athletes in the world, low confidence was like kryptonite, turning their cognition, emotion, and thoughts against them.

When our confidence is low, our toolbox shrinks. In the interviews conducted by Hays and colleagues, one athlete reported, "I was trying to use my psychological techniques . . . but none of them were working. I just couldn't concentrate. . . . Everything was going wrong and it was just horrible." A lack of confidence constricts our response.

As discomfort and doubt rise, the "devil on our shoulder" part of our mind goes on high alert, looking for evidence that backs up its view—anything that it can use to justify quitting or putting in less effort. When confidence is low, we are priming our minds to be susceptible to the negative spiral. We already have doubts over our ability to perform up to our expectations, so at the first sign of that being the case, our brain grasps hold of it. A gentle nudge and we're headed toward a full-blown freak-out.

Not surprisingly, when confidence is high, we experience the opposite. We're able to completely focus on the task at hand. We experience positive emotions: enjoyment, calm, and excitement. Our body language shifts, and we feel in control of the situation. Research shows we're able to cope with the demands of the situation, to frame nervousness as excitement, and to persist in the face of mounting fatigue. Compared to our low-confidence days, the cloudy and rainy day is exchanged for blue skies and sunshine.

Confidence is a filter, tinting how we see the challenges before us and our ability to handle them. It tips the scales toward an optimistic or pessimistic view of our current situation. When our confidence is high, we are able to cope with the demands of the event. We can manage our fears and doubts, quiet the negative voices, and redirect our focus to the task at hand. Confidence expands our ability to act, to manage, and to make our way through difficult situations. Confidence and toughness go hand in hand.

It's not a surprise that we see coaches, motivational speakers, and just about everyone in the self-help industry tout our deep need for belief in ourselves. We've long been told that through self-belief, we can accomplish anything we set our minds to. If we know the benefits of confidence, not only in allowing us to reach our potential but also in improving our well-being, why are even the best of the best filled with doubt? One reason: we've spent a long time developing the wrong kind of confidence.

When it comes to confidence, the old model of toughness

emphasizes acting instead of doing. We act confident, walking around with our chest puffed out, as though we have absolute belief and certainty in ourselves and our work. We talk a big game and never discuss our insecurities or doubts. It's all about the appearance of belief. But when push comes to shove, this external variety fails. True confidence has to be founded in reality, and it comes from the inside. It's not in ignoring the human condition of experiencing doubt and insecurity, but coming to terms with them and what you're capable of. It's not in the elimination of doubt, but in allowing enough doubt to keep us in check, while being secure in the knowledge that we'll find a way past the obstacle in our way. For far too long we've correctly insisted on the value of confidence, but we've gone along building the wrong kind.

"Fake it until you make it!" It's a well-worn piece of advice doled out to athletes, entrepreneurs, and anyone attempting to climb the corporate ladder. The advice encapsulates what we think of confidence: that it's essential, a requirement for success. And if we can't muster the real kind, we're better off acting like we know what we're doing than letting on the truth. But we didn't stop with advising adults. The search for an artificial form of confidence encompassed an entire generation of children as we espoused the benefits and virtues of self-esteem. Not by creating real value or through overcoming challenges, but by affirming to children how great they are.

TOUGHNESS MAXIM

Confidence is a filter, tinting how we see the challenges before us and our ability to handle them.

Building the Wrong Kind of Confidence

"Nice. Kind. Good friend. Fast runner. Penguin lover." Hanging on the side of my parents' fridge for the past thirty years, this sliver of laminated paper is a reminder of what my classmates thought of nine-year-old Steve. For my parents, it was a trinket, a sign that they were raising a "good" kid, whose peers thought highly of him. It was a nice gesture, a reminder of a kinder, gentler time. Though, to me, then and now, it meant something different.

That little laminated piece of paper with neatly penned handwriting was the result of an exercise from one of my favorite elementary school teachers. A task designed to improve our self-esteem, to teach us to be nice to one another, and generally to make us feel better about ourselves. We wrote down one compliment for each of our classmates, and then the teacher constructed our personal compliment paper. When I was nine, I remember feeling a bit strange about this activity. There were a few students in the class who weren't the kindest, who I had to search hard to find something nice about, and I generally left some sort of generic platitude as my response, like, "enjoys kickball." When I got my list back, I scanned it, distinguishing between those with truth behind them and those that had been written as some sort of default cliché. Some compliments had meaning; others didn't.

As a child growing up in the 1990s, I encountered a deluge of similar exercises aimed at enhancing my self-esteem. There were school-wide assemblies and classroom activities all aimed at making us feel better about ourselves. Away from the schoolyards, the evidence of this self-esteem movement could be seen in my trophy case—dozens of awards that signaled nothing more than that I had paid to be a part of a team. Win or lose, we all received a trophy.

I was a millennial who experienced the height of a psychological craze that had infiltrated the American mind, schools, and practice fields. The key to our childhood and societal ills had been found: a lack of self-esteem.

In 1986, California governor George Deukmejian signed legislation creating a task force that promised to change how we dealt with society's issues. The architect of the task force was John Vasconcellos, a California politician with a knack for the extravagant. Vasconcellos brought together two dozen experts from a variety of fields to tackle rising crime rates, drug abuse, declining educational standards, and a variety of other ills plaguing 1980s California. They formed the California Task Force to Promote Self-Esteem and Personal and Social Responsibility.

After undergoing self-esteem-focused therapy to help his own mental health, Vasconcellos transformed into an evangelist, proselytizing the benefits of self-esteem to all who would listen. His logic was straightforward: if we could make every person feel like they have worth and value, then each and every one could reach their full potential. If people didn't feel worthy, then it's no wonder they turned to drugs, alcohol, or criminal behavior. While his committee was initially mocked and ridiculed, Vasconcellos pursued his mission with religious fervor. He was going to change the world for the better.

Two years into their mission, Neil Smelser, the sociologist Vasconcellos had recruited to research the impact of self-esteem, gave a preliminary report informing the task force that "these correlational findings are really pretty positive, pretty compelling." Vasconcellos had found his proof—and sound bite. He plastered it across news stations far and wide, appearing on *The Oprah Winfrey Show* and *Today*.

In 1990, Vasconcellos produced his magnum opus, titled *Toward a State of Esteem*. In the executive summary, low self-esteem was declared a contributing factor to a slew of maladies, including drug and alcohol abuse, crime and violence, poverty and welfare dependency, and family and workplace problems. The 161-page report reads as if the group had found the key to fixing society. In fact, it states just that on page 21: "Self-esteem is the likeliest candidate for a social vaccine, something that empowers us to live responsibly and that inoculates us against the lures of crime, violence, substance abuse, teen pregnancy, child abuse, chronic welfare dependency, and educational failure."

There was one little problem. The conclusions were a lie, based on opinion, not what the research actually found. According to the data, the lone scientist, Smelser, concluded, "Self-esteem remains elusive because it is difficult to pinpoint scientifically. . . . The associations between self-esteem and its expected consequences are mixed, insignificant, or absent." The scientific validation was not there.

And that glowing sound bite by Smelser that landed the group on *Oprah*? It was taken out of context. When writer Will Storr went back and listened to a recording of the presentation given to the task force for a piece in *The Guardian*, he found the quote at the heart of the matter. Smelser's quote came when discussing a very small segment of the research on academic achievement before continuing with, "In other areas, the correlations don't seem to be so great, and we're not quite sure why. And we're not sure, when we have correlations, what the causes might be." His overall conclusion? Self-esteem didn't have much value.

It didn't matter that the research was inconclusive at best. The narrative was already written. Politicians and the media alike grabbed hold of the self-esteem movement and catapulted it into the stratosphere. Schools implemented the self-esteem interventions I and millions of others experienced as children. Even the way we talked to our kids changed. According to psychologist Jean Twenge, the frequency of slogans like "Believe in yourself and anything is possible" skyrocketed in the '80s and '90s. Posters with positive sayings adorned school classrooms throughout the nation. Before that? Not so much. As Twenge told *The Cut*, "They're all very individualistic, they're all very self-focused, they're also all delusional. 'Believe in yourself and anything is possible'? Nope, it's just not true."

As millennials, my generation often gets a bad rap. We're cited as self-centered, egotistical, and a bit narcissistic. We are overconfident, all believing that we should rocket up the success ladder, skip the mundane period of "paying our dues," and do what makes us happy. Every generation denigrates the next, but there is a kernel of truth to the complaints. According to research, millennials may have higher degrees of narcissism. They are more likely to see themselves as above average and have higher levels of self-esteem. But instead of complaining about the generation, we should ask ourselves: What else would you expect from a group that had it drilled into their heads that they were special and could succeed at anything they put their mind to?

Contingent Self-Worth and Seeking Self-Esteem

Feeling good about oneself has a natural appeal. It's easy to understand why a generation of parents, teachers, and administrators fell for the concept's importance. Self-esteem correlates well with a number of important factors related to health and well-being, including life satisfaction. Self-esteem is a good thing. But where we went wrong is thinking that self-esteem in and of itself should be the goal. That we should strive for the feeling, instead of having self-esteem be a by-product, something that occurs instead of is sought. The problem with the movement was that it put self-esteem as the focal point, something to seek out. Something that praise, ribbons, and rewards could cure. Self-esteem and confidence go hand in hand. They only work if they're founded in reality.

According to one prominent theory, self-esteem functions as a type of sensor, alerting us of our sense of worth or value. The sociometer theory posits that self-esteem represents a summary of our sense of acceptance, both from ourself and from our social group. The greater degree of acceptance, the greater our self-esteem. Successful people tend to have higher self-esteem not because they are striving for self-worth itself, but because it's a by-product of overcoming challenges and making meaningful connections with others. Our inner narrative changes when we are challenged and overcome adversity. When we put forth effort on a difficult task, we internalize that we have a strong work ethic. We become adept at knowing that we, too, can "grind away" at a problem. Lasting self-esteem doesn't come from being told that we are great. It comes from doing the actual work and making real connections.

With the self-esteem movement, we flipped the script, trying to give self-esteem without the accompanying action and work to validate it. Even worse, we shifted the focus away from the joy of actually doing the work and toward external praise and rewards. We were creating an artificial kind of self-esteem, a fragile one based on a delusion. We built self-esteem that was contingent and focused on the external.

When our self-worth is dependent on outside factors, we have what researchers call a contingent self-worth. We derive our sense of self from what people think and how we are judged. We give over control to external factors. When we utilize idle praise and combine that with undeserved rewards, we create an environment ripe for developing contingent self-worth. As Mark Freeman summarized in his book *You Are Not a Rock*, "The pursuit of self-esteem logically sets you up for low self-esteem. It's the same trap again: If you believe your value comes from people giving you things, then you hand over control of your self-image to other people. If they don't give you those things, then your brain logically concludes you must not be valuable."

Similarly, when our sense of self shifts to receiving praise or external rewards, our motivation shifts with it. In 2012, I was a young coach with a group of thirty willing volunteers at my disposal. Here was a group of college distance runners I referred to as my "ragtag team of misfits." They didn't fit the traditional mold of a college cross-country team. They were a mix of social and economic classes, ethnicities, and backgrounds that fit our inner-city university. Tasked with improving the group and recruiting better and better runners with limited resources, I set out to see if I could find clues on how to maximize performance and home in on selecting individuals who would thrive. I put my athletes through a myriad of psychological tests, evaluating everything from grit to mindsets to how they handled stress. There were some interesting tidbits, but the real value didn't come to fruition until much later.

Five years later, I'd largely forgotten about the data I'd collected. It sat in a file on my computer. But as I reflected on another season of performances, I decided to go back and calculate how much each athlete had improved. Some made giant leaps in performance, going from walk-ons to some of the best in the school's history. Others came in as phenoms and failed to live up to the hype. I was used to evaluating their lack of improvement through a training lens. Either the training worked or it didn't, the logic went. But as I peered through the variety of improvement curves, the psychological data I'd collected long ago came to mind.

When I went back and compared motivation styles to performance improvement over each athlete's career, one factor stood out. Those who scored high in a particular type of extrinsic motivation called external regulation had lower improvement rates. External regulation is defined as when "the sport is performed not for fun but to obtain rewards (e.g., praise) or to avoid negative consequences (e.g., criticisms from parents)." The five highest-ranked athletes in external regulation were five athletes who showed the least amount of improvement. As we were reading out the names, my assistant coach Nate Pineda blurted out, "This is crazy! What a correlation." We were taken aback by the names at the top of the list. They were those we struggled as coaches to figure out how to help improve.

When it came to self-esteem, we tried to boost a generation with methods that pushed us toward having the contingent and externally driven kind. When self-esteem is contingent, it's fragile. When it's based on external rewards or praise, it's dependent on something over which we have little control. When it comes to developing confidence, we often fall to the same fate. We attempt to create a fragile sense of confidence, one based on bravado and external displays. We try to prop up our confidence levels with praise, rewards, and even grade inflation, which often is unearned or undeserved. We believe that failure of any kind should be avoided instead of embraced, because it shatters our confidence. We're making the same mistakes, setting ourselves up for a confidence based on the external, not internal.

Arrogance Sits on Insecurity; Confidence Sits on Experience

On the ESPN program *Always Late with Katie Nolan*, Nolan had a college student named Darrell participate in a storied football drill, the 40-yard dash. Darrell wasn't a lucky contestant; he was on the show because of his brashness, tweeting, "I really want to know how I would fare in the 40-yard dash. I'm confident I could run a 4.4 time." Nolan challenged Darrell, an avid gym rat, to put his money where his mouth was. Before he laced up his shoes, Nolan asked, "Confident you can run a 4.4 time? You're Odell Beckham Jr., that's what you think?" Darrell replied, "One hundred percent." Darrell didn't exactly sprint his way to a 4.4. Using a stride that resembled more of a shuffle than a sprint, he clocked 5.5 for his dash. A time that would put him more in line with the slowest of the slow 300-pound linemen. Or in a completely different universe from his goal.

Confidence is something that we intuitively understand. It's a feeling that makes us think, "I've got this." When we think of tough individuals, we often picture those beaming with confidence. But what we're often left with isn't a secure and substantiated type of confidence, but one that resembles our 40-yard-dash friend Darrell's, a fake kind.

Just like self-esteem, with confidence, there's a real version—one that is deep, based on evidence and understanding—and a fake version that's based on bravado. The fake version is derived from insecurity. It's a mask that a person wears, attempting to fake his way through a task or to put on a show for his friends. Men seem to be more susceptible to the fake variety, perhaps best demonstrated by the fact that one in eight men somehow think they could score a point against tennis superstar Serena Williams. Delusion and fake confidence go hand in hand.

This brashness or arrogance is a confusing phenomenon, one that transcends sports. No one exemplifies this more than former president Donald Trump, who has declared a long list of topics that he "knows more about than anyone," including campaign finances, courts, social media, renewable energy, taxes, construction, technology, and dozens of other items. Being loud and boisterous doesn't signal confidence. It's the opposite. Recent research found that those who tend to shout the loudest—both in person and online—do so because they lack inner confidence. We often confuse arrogance and brashness with confidence. We mistake external displays for indicators of their inner workings, not realizing that the need to proclaim that one is confident is undoubtedly a sign that they are anything but.

We even encourage this conceptualization. We've demonized doubt. Showing any weakness, having any hesitation, is a sure sign that you don't deserve the raise. Humility and vulnerability are signs that you can't handle "tough" situations. It's a message we've absorbed since peewee football. Acting confident leads to success. Let any doubt or uncertainty enter your mind and you are on the wrong side of the equation, heading toward failure and breakdown.

We tell each other to fake it until we make it. Or, after a poor performance, we tell athletes to act confident, as if it's something that they can simply switch on. We've confused the outward displays with an inner confidence. We think that if we can talk the talk, we'll be able to walk the walk. Just as with self-esteem, we've gotten it wrong. Confidence has to come from deep within.

When we face a challenge, expectation and reality should have a high degree of overlap. Whenever they do not (i.e., a high sense of confidence and low ability to meet the task's demands, or vice versa), then our likelihood to persist through a challenge or even perform at our best is greatly diminished. We're more likely to choose the easy way out. To stop. To quit. To find a solution that doesn't involve confronting whatever is in our way. Why would we want to set ourselves up with an unattainable inflated expectation, only to have reality smack us in the face when we enter the arena?

When your bite can't back up your bark, your brain is already aware. It's no dummy; it has a job to protect you instead of letting your foolish ego get in the way. When we are overconfident, we set ourselves up for failure. This isn't idle conjecture: researchers have found this phenomenon in everything from competing in sports to deciding whether to stay in a relationship or quit your job. It's easy to feel confident in the beginning but when we come face-to-face with the reality that we might fall short of our goal, we experience what psychologists call an action crisis. We shift from a goal-directed orientation, where the motivation to achieve is the focus, to a state where negative thoughts and sensations persist. We shift from being driven to succeed to negotiating with ourself to abandon the goal. False confidence sways that vital balance point of our capabilities versus the difficulty of the task. If we overestimate our abilities, then at the first moment that our mind realizes, "Wait a minute, we are in over our head," it hits the protect button. We shut down, see the task as a threat, and save our energy for something more worthwhile.

For example, take the students who go into a test acting confident, even though they did the bare minimum of studying to prepare for it. Sure, they can fake it, pump themselves up, convince themselves they can rely on their ability to BS on the writing prompt, or use the multiple-choice test to their advantage. But the moment they encounter a few questions that leave them feeling hopeless, reality seeps in, their body is flooded with stress hormones, and panic takes over their minds. The greater the mismatch between expectations and reality, the worse off we are.

Ming Ming Chiu, a professor at the State University of New

York at Buffalo, set out to evaluate the impact of confidence on children's reading levels. In looking at students across thirty-four countries, Chiu and his colleagues found that while a little confidence might help, too much can be detrimental. Overconfidence was linked to worse reading comprehension. In explaining the findings, Chiu reported, "If an overconfident student chooses a book that is too hard—such as *The Lord of the Rings* rather than *Harry Potter and the Sorcerer's Stone*—he or she might stop reading after a few pages and let it sit on a bookshelf. In contrast, a more self-aware student is more likely to finish an easier book and continue reading more books."

If we artificially elevate our confidence, telling ourselves this will be a piece of cake or we've got this in the bag, our brain is constantly receiving the message that we won't have to work hard to achieve our goal. If it's supposed to be easy, why should we waste excess resources? When reality hits us, we jump straight to a freakout. "What's happening?! I thought this was going to be easy or within our capabilities," our mind might think. Cultivating fake confidence creates insecurity for our minds to exploit.

This is the difference between a fragile outer confidence and a secure inner one. In a study on more than twelve thousand individuals, researchers found that faking it might help a touch when compared to those who lack any confidence. But when compared to those who had inner confidence, it failed. In ratings of self-motivation, self-esteem, resilience, coping skills, adaptability, and assertiveness, being high in outer confidence may have led to a tiny improvement in the aforementioned measures, say from a 35 out of 100 to a 42. But those who possessed a high level of secure inner confidence were scoring in the 70s and 80s on the same measures. Inner confidence leads to meaningful change. As Ilona Jerabek, president of the company involved in the research, commented in reflecting on the study results,

"Pretending to be confident can be effective to some degree . . . however, like any façade we create, it won't last."

False confidence helps in situations where we largely don't need an extra boost. Faking it works on easy tasks, where the challenge is low and a bit of extra motivation is needed to get you started. In the workplace, research shows false confidence can fool those who are uninformed on a subject, but those with even a moderate understanding of the topic will sniff out your lack of acumen. In situations that demand toughness, false confidence largely fails. Outer confidence is fragile, falling away when pressure or uncertainty arises. A secure inner confidence is robust. While we envision tough competitors and executives as having an unshakable belief in themselves, the reality is that the best way to be prepared for a challenge isn't bravado but tragic optimism, a sense of reality in the short run but hope over the long haul.

How to Create Inner Confidence

We've gone about creating confidence all wrong, thinking it must mean that we are confident in all situations regardless of our capacity to face the challenge. That the tough can take on anything. We need to lower the bar and realize that confidence simply means having security in knowing that you can accomplish whatever is within your capabilities. It's not in being able to do the impossible. To develop true, inner confidence, there are four steps:

1. Lower the bar. Raise the floor.
2. Shed perfection. Embrace who you are.
3. Trust your training. Trust yourself.
4. Develop a quiet ego.

1. Lower the bar. Raise the floor.
When trying to improve, most of us go for the lift-the-ceiling approach, judging ourselves by our best performance ever. In track, we would define ourselves by our personal best for an event. To get better meant running faster than we ever had before. Brian Barraza, a professional runner, sees performance in a different light. "My goal is to raise the floor. Being confident in that whenever I step out on the track, I'm going to be able to run a certain time." Instead of going all in for the massive breakthrough, Barraza sets a minimum expectation. "When you raise the floor, it allows for those days when everything is clicking to exceed expectations. It's not that we are lowering our ceiling or playing it safe; it's that we've developed the confidence to know that X performance is repeatable. That as long as we do what's in our control, we can achieve a certain standard, no matter the circumstances," Barraza told me one day after practice. As I watched this idea percolate through the athletes, I came to notice a trend. Those who raised their floors had an inner confidence about them. What once seemed crazy to contemplate was now the norm.

Brian Zuleger, a sports psychologist out of Adams State University, taught me an exercise to reframe expectations. Instead of aiming for our best performance, something that we can only accomplish rarely, shoot for improving your best average. When we judge ourselves against our all-time best, we inevitably fall short more often than not. Instead, averaging out our five most recent performances gives us a still tricky but achievable goal.

The aim is first to be consistent. Don't lower your expectations just so you can become confident. Understand what you are capable of, and set a standard that falls within that realm or just a touch outside of it. Embrace reality. Understand that a breakthrough doesn't come from creating a false sense of confidence; by developing the belief that you can achieve a certain standard, you free yourself up to take risks when the opportunity presents itself.

2. Shed perfection. Embrace who you are.

Real confidence lies in understanding who you are and what you are capable of. It lies in being vulnerable, not in delusional machismo. You don't raise your floor by developing an unrealistic view of yourself. You do so by taking a hard look at where you are in the moment. Understanding what you are capable of, what challenges the task brings, and where your weaknesses might lie. Real toughness resides in being humble and wise enough to acknowledge your strengths and weaknesses. To find the right point of risk versus reward, to balance upon the expectations-versus-demands scale. Vulnerability—in acknowledging that you aren't going to be stoic, be impervious to pain or pressure, or never fail—is the only way to obtain inner confidence. Truly tough individuals don't mind exploring their weaknesses. They develop the capacity to express vulnerability and pain without fear of being shamed. Refusal to explore or acknowledge your weaknesses is a sign of insecurity, not confidence.

When we are vulnerable, the words, phrases, and criticisms that might get under our skin lose their power if we've acknowledged and come to terms with them. It's when we try to hide our insecurities that they can be exploited. Developing fake confidence is a form of masking—a delusion to try to fool ourselves into thinking we have what it takes. We create fake confidence for the same reason we build fake self-esteem: to protect the sensitive parts of our ego and to hide our weaknesses and insecurities from the world for fear of being exposed as a fraud or as not good enough.

If we can come to terms with who we are, warts and all, we slowly disarm our insecurities, such as our tendency to wince when we get criticized about our looks, writing prowess, or intelligence. We can reformulate our relationships with these items, treating them not as things to mask or hide but as items to know and learn from.

When we come to terms with our shortcomings, we're able to adopt a secure sense of self. In this chapter, we've discussed the human battle with two major psychological constructs: self-esteem and confidence. Both come down to the same issue: how we see ourselves and the world around us. In other words, do we see ourselves with clarity or delusion? Our sense of identity plays a large role in how we see the world. When we're young, we try on different identities, dabbling between musician, jock, nerd, or whatever other labels teens utilize. We can ditch one for another in as short a time as our winter break. As we age, our identity begins to cement. But how much it hardens is up to us.

We may not want our chameleonlike identity of our preteen years, but who we are shouldn't be set in stone either. We need to feel comfortable with who we are but be able to change. We need a secure but flexible identity. If our identity becomes set in stone, then any sort of threat to our inner narrative is interpreted as an attack. We'll dig in, defending who we are at all costs and utilizing our cognitive bias to keep our sense of self intact. If, on the other hand, we acknowledge our foibles, then we can take ourselves out of this defensive stance, with the knowledge that the structure of who we are is stable, but the details are up for revision. It's not an attack, but an aid to understanding our weaknesses.

3. Trust your training. Trust yourself.
At the beginning of this chapter, we discussed coach Fred Wilt admonishing marathon runner Buddy Edelen for his insecurity. Wilt's message was a simple one that's been repeated by coaches for generations: trust your training, trust your fitness. These simple phrases are meant to relay a much more profound lesson: that true confidence is founded in doing the work.

Confidence doesn't come from doing the work out of fear or neuroticism—to practice because you are afraid to lose or fail. When fear drives the motivational ship, then insecurity pervades. When the work is done in the name of getting better, of enjoying the process, of searching for mastery of the craft, then confidence gradually grows. A feeling that "I've been here before, I'm prepared for this challenge." It's the writer who shows up at his desk every day and writes. The dancer who spends countless hours perfecting her routine. The executive who game-planned just about every scenario possible. The result might not always end up as hoped for, but doing the work provides a secure confidence founded in something concrete.

LeBron James is arguably the best basketball player on the planet. In 2014, he added a new element to practice: a one-on-one contest where buckets only counted if you shot off the wrong foot and with the wrong hand. James was minimizing a weak point in his game. Shooting off the wrong hand and foot would never come as natural as his dominant side, but it allowed him to be just competent enough in case he was put in that situation in a game. He gave himself a chance.

It's no wonder that his counterpart in the argument for best player in history, Michael Jordan, once said, "If you have doubt or concern about a shot, or feel the 'pressure' of that shot, it's because you haven't practiced it enough. The only way to relieve that pressure is to build your fundamentals, practice them over and over, so when the game breaks down, you can handle anything that transpires." To gain confidence, put in the work from a place of growth, not fear. Boldness is earned, not assumed.

4. Develop a quiet ego.

Our ego is like a kid trying to fit in during middle school. It just wants to be liked. The moment it feels like failure or embarrassment is on the horizon, it quickly finds an out, diverting responsibility and distancing itself from the situation. It's the child who tries to blame his younger sibling for the spilled milk on the carpet, or finds any excuse—"The teacher has it out for me!"—whenever he brings home a history test with a bright-red *F* on it. The ego is all about protection.

We like to walk around with a story in our head that we are a good, decent, competent person. Whenever evidence presents itself to the contrary, our ego goes into overdrive to rationalize, justify, or explain away why the opposite cannot, must not, be true. Our ego does many good things for us, acting like a social immune system that swats away psychological threats. But if it is overactive, propping up a sense of self that doesn't reflect reality, then it's just as damaging as a hyperactive immune system. We don't want to shut off our ego. We just want to dampen it down to a reasonable level.

Social psychologist Heidi Wayment has pioneered the idea of a quiet ego. As she told *Scientific American*, with a quiet ego, "the volume of the ego is turned down so that it might listen to others as well as the self in an effort to approach life more humanely and compassionately."

A quiet ego is about keeping ourself in balance—coming to

terms with the need for confidence, but being keenly aware of the strengths and weaknesses of ourselves and our situation. It's being open and receptive to others, instead of defensive and closed off. It's having the ability to zoom out, gain perspective, and understand that a short-term loss is often part of a long-term gain. How do we quiet our ego? Ask: What causes you to sting, ruminate, and pull away? What causes you to default toward defensiveness? Do you dismiss criticism out of hand, or do you consider and evaluate it? What you're after is a dash of self-awareness and reflection combined with a secure sense of who you are. A bit of doubt and insecurity is normal. Too much defensiveness and protection are signs your ego's too loud. When we mix perception, awareness, and security together, we can move on from the false-bravado style of confidence that permeates the world. Confidence is doing difficult things, sometimes failing, but seeing where you lie, and then going back to the work.

***

What do we do when we fail? If you get an F in math class, do you now think that you are bad at math? That it's not your thing? If you exceed your earnings goal for the quarterly report, do you attribute it to your cunning business expertise? A large part of developing confidence lies in creating a secure but flexible sense of self. And a large part of that depends on how we integrate success and failure into our inner story.

In her PhD thesis, sports psychologist Jennifer Meggs at Teesside University found that we generally assimilate positive and negative beliefs into our sense of self in two different ways, either compartmentalization or evaluative integration. When we compartmentalize, it's all or nothing. We see the item as either entirely positive or entirely negative. Take the example of failing in a class at school: compartmentalization tells us that it's all bad news and that we probably don't have a future in the field.

On the other hand, those who possess an evaluative integration bring more nuance to the discussion. They can see the good and bad in situations. They might feel anxious or frustrated, yet still believe that they can perform the task at hand. It isn't all or nothing. When researchers tested individuals based on what self-structured group they fell into and compared that to their scores on a toughness scale, the results were clear. Those who were better able to integrate instead of compartmentalize were tougher and better at "thriving in adverse circumstances."

And that's the key to true confidence. Acknowledging the good and bad, our weaknesses and strengths. Living with and dealing with reality instead of putting on a front. Setting our own standards. And realizing that, as Alain de Botton said in his book *On Confidence*, "The way to greater confidence is not to reassure ourselves of our own dignity; it's to come to peace with our inevitable ridiculousness."

CHAPTER 5

Know When to Hold 'Em and When to Fold 'Em

In Russian physiologist Ivan Pavlov's original experiments on classical conditioning, he ended up conditioning dogs to salivate when a bell was rung. Over time, with the ringing of a bell at mealtime, the dogs connected the noise with the notion that food was on its way. Pair a stimulus with a positive response, and you get dogs drooling in anticipation of food once that bell is rung, even if it never comes. Decades later, Richard Solomon's lab at the University of Pennsylvania was investigating a close cousin to classical conditioning, only with a negative twist. Instead of tying the ringing of a bell to food being on its way, the playing of a tone was associated with receiving an electric shock. They were testing fear conditioning. A sound was played; the dogs were shocked. Over and over until the tone became a warning sign that a shock was on its way. Twenty-four hours later, the conditioned dogs were placed in a box with only a short barrier keeping them in. It was easily escapable. Solomon's lab believed with fear conditioning the dogs would try to run, jump, or do whatever it took to get away once they heard the tone. A shock was on the way, after all. Except the dogs didn't try to escape. The tone was played, and the dogs just sat there.

Martin Seligman and Steven Maier were two young graduate students who had just joined Solomon's lab. They were convinced that "a profound failure to escape was *the* phenomenon" to study.

Perplexed by the apathy of dogs who knew a tone meant a painful shock, Seligman and Maier sought to figure out what was going on. In a series of experiments, they again delivered shocks to dogs, but this time instead of being trapped, they gave some of them a way out. For half of the dogs, if they wanted to avoid the pain, to turn the shock off, all they had to do was press their heads against a panel. These dogs quickly learned how to end the experiment. For the other half, the panel didn't work. They had to sit there, unable to control when the shocks would start or end. They experienced seemingly random shocks throughout their testing time. One group of dogs could escape. The others were trapped and had to endure.

Following their experience in the shock apparatus, both sets of dogs were put inside a cage consisting of two rectangular compartments separated by a small wooden barrier only a couple of inches high. The cruel game of shocks and avoidance wasn't over. This time, a shock was delivered through the floor to one compartment, but not the other. Unlike the first go-round, to avoid the shock, every dog had the same opportunity. Hop over the small wooden barrier to the other chamber, and pain was avoided. For all, safety was just a step away.

What Seligman and Maier found was astonishing. The dogs that learned to escape shocks in the initial experiment by pressing a panel did the sensible thing in the latter experiment: they jumped the barrier to safety. The dogs that had no way to end the shocks in the initial experiment? Despite having the same opportunity and capability to hop to safety, they didn't. Over two-thirds of the dogs didn't even attempt to find an escape, even after ten trials. And the ones who eventually figured out they could escape took nearly until the end of the experiment, where their counterparts figured it out on the first shock. The poor dogs who struggled to escape became listless, whining and cowering in the corner. They had lost the ability to try.

Seligman and Maier stumbled upon a phenomenon they called learned helplessness. The dogs had learned that pain and suffering were outside of their control. They had no power over what was happening to them, so their only point of recourse was to sit there and take it. The same phenomenon has been replicated in other animals. When researchers shocked rats the exact same amount, those who could not press a lever to escape demonstrated a greater stress response and double the amount of gastric ulceration. When the rats lacked control over their fate, they mimicked the dogs: they stopped trying to escape their situation. The pattern was clear. Take away control, and animals resign themselves to their fate. They give up. Even if the path to avoiding despair is right in front of them.

Our modern workplaces, sport leagues, and even schools often train us to respond in the same manner. They, like the old model of toughness, often rely on control and constraint. They take choice away. It's the dictatorial coach who motivates through fear and punishment. The boss who micromanages. The company that tracks every minute that a worker is on task, and when they click away to Facebook. The parent who restricts and controls their child so much that they cut off their natural inclination to exploration. It turns out, when control and choice are taken away from humans, we act just like the helpless dogs in the experiment.

We lose the ability to try. Lack of control extinguishes the flame of even the most motivated. When we lack control, when we feel like no matter what we do, it doesn't make any difference, our brain is getting the message "What's the point?" In our modern workplace, we see the transformation before our eyes. The young and enthusiastic junior staffer gradually transforms into the mindless occupant of a cubicle. We move from thriving to surviving. There's a reason that burnout is rampant in just about every profession. We have spent years training hopelessness in the misguided name of discipline. A new approach is needed, one that is based on giving back and bolstering autonomy. One that allows us to unlock the tools to navigate through discomfort.

Give-Up-Itis

In 1606, 105 men set sail aboard the *Susan Constant*, *Godspeed*, and *Discovery* to traverse the Atlantic Ocean and establish Jamestown, the first permanent English colony in the New World. Surviving a four-month journey, the mix of gentlemen, blacksmiths, carpenters, and all-around handymen established their new home on the banks of the James River on May 13, 1607. Given the failed expeditions that preceded the Jamestown colonists and the inherent risk and danger of traveling to a still-unexplored land across a vast sea, the adventurers were well aware of the danger before them.

At the end of the first year, only thirty-eight of the original settlers remained alive. Despite the astronomical death rates, colonists kept coming. As a colony, Jamestown faced long odds. They experienced severe drought, difficulty in establishing crops, infiltration of disease-spreading mosquitoes, and hostile situations with Native Americans. Despair, isolation, starvation, and death all were commonplace.

In letters sent back home, colonists reported lethargy and apathy spreading throughout the settlement. In a letter from 1610, William Strachey writes of extreme idleness, an inability of settlers to "sow corn for their own bellies, nor to put a root, herb, etc. for their own particular good in their gardens or elsewhere . . ." While another colonist reported a few years later that "most give them selves over, and die of Melancholye."

Looking back on this harsh period, the common explanation

for the constant reports of apathy and laziness are either that starvation sapped colonists of any energy and desire for work or that the colonists were unprepared for the harsh realities of the New World. Modern historian Karen Kupperman has a different take. While starvation was a real and constant threat, what if the psychobiological toll of living with death and despair as constant partners led to death itself?

As Kupperman outlined in her paper "Apathy and Death in Early Jamestown," there was a brief respite in the astronomical death rates of early Jamestown. When John Smith, adventurer and later legend of the Pocahontas tale, took charge with a new motto, "Work or starve," deaths briefly subsided. Requiring each settler to put in four hours of farming per day, Smith later reported that only seven or eight men died during his ten months in charge, a drastically lower rate than before or after. Smith made the case that "idleness and carelessness" were the real cause for concern in Jamestown. Others seemed to agree. In 1620, colonist George Thorpe reported, "More doe die here of disease of theire minde then of their body." Was the simple desire to live not enough to motivate the Jamestown settlers to lift themselves up and farm for the sake of their own survival? Or were they helpless?

In the Korean War, prisoners of war experienced a new disease. The patients succumbed to a sensation of listlessness, as if emotion and life itself had been drained out of them. Over time, they would fade, forgoing their daily routines, and giving up on their hygiene. Even their movements took on a zombie-like pattern, feet lifting off the ground as minimally as possible, as if to conserve every ounce of energy. The disease had no signs of organ failure or identifiable internal distress, but the prognosis was grim: death. The disease was give-up-itis.

Give-up-itis didn't originate in the Korean War. Similar reports abound from those who face extreme traumatic experiences. During World War II, those who suffered through concentration camps reported comparable experiences, tales of friends whose inner light would slowly fade as they transformed into listless beings. More recently, individuals who were lost at sea report shipmates who didn't survive the journey experiencing a similar phenomenon.

In 2018, psychologist John Leach tracked the history and science of this phenomenon in a report entitled "Give-Up-Itis Revisited." According to Leach, sufferers of give-up-itis progress through a series of five stages. It starts with a general withdrawal before turning into apathy, loss of emotional response, and a lack of response to any external stimulus. Along the way, motivation to do menial tasks erodes until the simplest of tasks can no longer be completed. Listlessness takes over. The last stage is psychogenic death.

The strange case of give-up-itis is easy to dismiss. Dying of a lack of motivation to continue living doesn't fit well with our medical minds. We need a cause, a failure of an organ, a disease that spreads. But we saw the same phenomenon in dogs who were put through shocks and torment—a descent into a passive state where escape is unthinkable. And in other research, rats conditioned for learned helplessness gave up and drowned within minutes when placed in a water tank, even though they'd demonstrated the ability to swim for much longer, some for hours on end. Give-up-itis might not always lead to death, but the feelings of apathy when we lack control are real and frequent. In a sad twist, the CIA incorporated these findings in the brutal interrogation techniques, such as waterboarding, that were utilized in the early 2000s. According to a Senate report, the CIA hoped to eliminate any "sense of control and predictability" to create learned helplessness and apathy.

There may be a cure for give-up-itis. Psychiatrist and Holocaust survivor Viktor Frankl noted that when he was in Auschwitz, another prisoner told him that to increase his chances of survival he should do two simple things: shave and stand tall. In other words, control what you can. Not too dissimilar from the advice that John Smith gave to his Jamestown colleagues. Leach believes that bringing some sort of normalcy to perilous situations "requires an appraisal that the person has, at least, some control over his situation, has not accepted mental defeat." He goes on to conclude, "That sensation of choice indicates a reversal of mental defeat and the reimposition of some personal control over the situation which is a key factor in recovery. . . . [Give-up-itis] is the clinical expression of mental defeat; in particular, it is a pathology of a normal, passive coping response."

Give-up-itis isn't restricted to those moving to distant lands or experiencing extreme acts of survival; in many ways it's the scourge of modernity. No, death isn't the end result here, but listlessness sure is. Work is no longer nine to five. Emails have to be answered at all times. We are constantly on, constantly connected. If you don't respond to a parent, boss, or customer at nearly any hour of the day, you're in trouble. We're expected to "grind" endlessly in our pursuit. And if we feel overwhelmed, there are dozens of others lined up for a shot at our job. In the modern workplace, we give away any sense of autonomy in exchange for the right to grind for a few extra bucks. As work infiltrates all aspects of life, we are actively training learned helplessness. When we feel like we have no say or choice, we're on the path toward listlessness. We are the dogs who learned that no matter how many times we press the button for help, no one is coming. It's no wonder difficult decisions now seem overwhelming. Or why many of us struggle to get out the door to exercise, or even do the dishes at night. We blame it on lack of will, or motivation, but the truth is when we lack a sense of control over our life, apathy naturally takes over.

Choosing to Be Tough

After receiving his MD from Emory University in 1962, Peter Bourne served in the US Army for three years during the Vietnam War, a conflict that epitomized the terrors of war in an environment that was unlike other recent conflicts. Surprise and ambush replaced the structured wars of years past. For someone with a keen interest in how individuals responded to stress, Vietnam served as the perfect place for Bourne to understand how the body reacts to such perils. With their camp situated on the Ho Chi Minh trail, a direct supply route for the Vietcong, Bourne began taking daily measurements of the stress hormone cortisol in a handful of soldiers, officers, and radio operators, in anticipation of an impending attack. Bourne was trying to understand how the body anticipated and responded to stress, but also how it recovered. Once that inevitable attack finally came, Bourne continued tracking stress hormone levels as the fighting subsided and normalcy returned.

When he compared the cortisol response among the different groups, a pattern emerged. The soldiers all saw their cortisol levels drop significantly on the day of the expected attack, while the radiomen and the officers experienced the opposite, an increase. After the attack, the soldiers' cortisol levels increased, while the officers' decreased, both returning to baseline. The radio operators remained elevated even after the attack had concluded. All the men were in the same camp, experiencing the same environment, expecting the same attack, yet they had different stress responses. Bourne believed that the difference lay in the level of perceived control.

Through a combination of training and circumstances, the soldiers were able to cope with the impending raid. Fighting was what they had prepared for. They'd been drilled and drilled on what to do. Their job wasn't to design the plan or develop tactics.

It was to follow orders. And they had an inherent belief that if they followed their instructions, they would succeed. They were the best-trained army in the world, after all. They didn't have time to ruminate on the issue. Like John Smith's Jamestown colonists centuries earlier, the soldiers had tasks to complete. They had to prepare their camp, ensure their defensive perimeter was stout, and prepare their weapons. Thanks to training, they felt like they had a degree of control over a hazardous situation.

Officers didn't have this luxury. It may seem like they would have more control, but when it came to impacting the potential outcome, the officers had less. They had more information and intelligence than the soldiers on the ground. But they received orders from higher command, and then made their best guesses on strategy and tactics, and hoped it would work. Similarly, the radiomen were reporters of information. They were the go-betweens: important, but they had little say in or impact on what was going on out on the battlefield.

Our level of control changes how we respond to stress. Not just on the battlefield, but in every aspect of our life. Think: the middle manager who feels trapped between her boss's dictates and the workers' needs, or the teacher who is told not to deviate from the approved standardized curriculum, even though their students are falling further and further behind. When we lack control, our stress spikes. When we have a sense that we can impact the situation, our cortisol response is dampened. Control doesn't alter just our hormonal response but also the experience that accompanies stress. When researchers peered into the brains of subjects with fMRI machines, they found that when pain was controllable, participants had lower rates of anxiety, along with a decreased response in the threat-sensing area of the brain (amygdala). They had not only a lower alarm response from the amygdala, but also a better-equipped controller (the prefrontal cortex) that was able to step in and put out the fire much more quickly. When we have a sense of control, our alarm is quieter and easier to shut off.

Control alters not only our physiological response to stress, but also our ability to persist. When we believe we have influence over an outcome, we're more likely to persevere, even if we face a setback. The traditional test of endurance capacity is a VO_2 max test. It's a cruel task that's been a staple of exercise physiology for a century. Run on a treadmill, as the physiologist increases the speed or incline every minute or two until you either choose to stop or fly off the back of the treadmill from pure exhaustion, all while breathing through a mask-and-tube contraption that measures how robust and efficient your cardiovascular system is.

I'd performed several such tests in my running career, but something about them always bothered me. They bear little resemblance to an actual race. When competing out on the track or road, you are in charge of when you speed up or slow down. In the VO_2 max test, that choice is taken away. So one day in the fall of 2014, I hopped on the treadmill, determined to do a different kind of test, one that gave the control back to the athlete. In the new test, the goal wasn't to hang on until I cried "uncle"; it was to exhaust myself in about ten minutes. I was in charge of pacing, able to increase or decrease the speed or incline as I saw fit. When I hopped on the treadmill, the dread was gone. I was familiar with bringing my body to exhaustion, especially when I knew about where the finish line was. Upon completing the test, not only did I enjoy it more, but how much oxygen my body could take in and utilize was significantly higher.

The next step was an obvious one: ask some athletes I coached if they'd like to be guinea pigs. Over the following week, I put them through two VO_2 max tests, the traditional one where control was in my hands, and the new version where they were in control. When we crunched the data, the athletes who were faster on the track did better in the new style, reaching a higher VO_2 max when they had control. On the other hand, those who weren't quite as fast on the track were slightly better in the traditional test.

In discussing with the athletes afterward, it was clear the better runners needed freedom. They wanted to explore their limits and be in charge of how they did. For the not-quite-as-fast runners, reducing the task to either persist or quit simplified their goal. They could maintain focus on one thing, to just keep going, which helped their performance. For the more experienced and better runners, this wasn't enough. They needed to be in charge, to be able to make a choice to run their best.

TOUGHNESS MAXIM

Our level of control changes how we respond to stress. When we have a sense of control, our alarm is quieter and easier to shut off.

The Science of Choice

The ability to choose is not just related to persistence and performance. It is deeply ingrained and required for being a normally functioning human being. The two most prominent theories related to motivation and human flourishing are self-determination theory and self-efficacy. According to self-determination theory, the level of autonomy, or "the desire to be causal agents of one's own life," is intricately tied to our well-being. It serves as one of the three basic psychological needs that allow us to flourish and bolster our motivation. In legendary psychologist Albert Bandura's seminal theory of self-efficacy, control plays a similarly large role. According to Bandura, self-efficacy "reflects confidence in the ability to exert control over one's own motivation, behavior, and social environment." Both theories are tied to motivation, work performance, happiness, well-being, life satisfaction, and academic success. When we feel like we can have an impact on whatever it is we do, we are better off. The ability to have control is central not only to overcoming adversity, but also to being a happy, healthy human being. And it's reflected in our brains.

When we are given a choice, our brain responds as if having a choice is the reward in and of itself. The striatum, an area linked to reward processing, activates when we have the ability to choose. In the lab, when we are given a reward for choosing the right answer, our striatum lights up. When we are given that same reward based on luck or chance, the striatum remains silent. The ability to choose improves performance not only in athletic tasks, but also in everyday ones. Giving nursing home residents more autonomy and choice over their care and their surroundings improves mood, alertness, and well-being. While in the workplace, those who report feeling more autonomy and less micromanaging have higher levels of job satisfaction and performance. We have a deep need to be in control over our environment and, in particular, our lives. When we give away this sensation, we lose a sense of ourself. And if we repeatedly do so, we lose our ability to respond even to the simplest of challenges.

In 2011, when Steven Maier, the scientist who pioneered the work in learned helplessness, went back to utilize modern neuroscience to expand on his seminal work, he made a subtle adjustment to his original theory. Learned helplessness implied that the dogs were learning not to try. That after suffering and despair, they'd given up. Maier and colleagues evaluated rats using a similar shock paradigm to the one they'd used in the 1960s, with some receiving controllable and others uncontrollable shocks. But this time, after going through the shock protocol, they put all the rats in a cage with a juvenile rat while monitoring their brains. When new rats encounter each other, they act similar to dogs, the dominant or older one sniffing the younger one. The rats that received shocks that they could stop or control followed the typical pattern, sniffing their younger counterpart. The rats that received uncontrollable shocks? They cowered in the corner, barely acknowledging the juvenile rat.

When Maier peered inside the rats' brains, he noticed a distinct difference. The dorsal raphe nucleus (DRN) area in the brain lit up and stayed activated in the cowering rats. The DRN is a primitive brain area that responds to stress and releases serotonin into adjacent areas in the brain. Their counterparts who could stop the shocks also experienced an initial stress response, but an area in their prefrontal cortex soon lit up, shutting down the DRN's primitive stress response. If you remember from earlier in the chapter, the prefrontal cortex acts as a control, signaling to dampen down the alarm. Maier took it a step further, stimulating the prefrontal cortex in the rats that had an overactive stress (DRN) response thanks to their uncontrollable shocks. All of a sudden, the rats transformed, no longer displaying any characteristics associated with learned helplessness. The alarm had been shut off, quieted down. As Maier told the American Psychological Association, "It's like the forebrain is saying, 'Cool it, brain stem, we have the situation under control.'"

The rats and dogs weren't learning how to be helpless. That was their default state. The dogs had to learn that they had control. That they could, in scientific terms, activate their prefrontal cortex. Turn on their controller, which allowed them to turn the alarm off. If they felt like they had control over the situation, they wouldn't succumb to the listless apathy that lack of control seems to foretell. We need to train to be able to turn the alarm off. It wasn't learned helplessness. It was learned hopefulness.

When life feels like it's spinning out of control, or like the task you have in front of you is insurmountable, it's easy to default to hopelessness. To "What's the point?" That's natural. Your body evolved to conserve energy. We need to train hopefulness. To clear the path to continue. It doesn't take big heroic efforts to train hope. Small signals that you are in control, that you can have an impact, will be enough to turn our prefrontal cortex back on. If too many emails are causing you consternation, define a specific hour each day in which you'll answer them. If grief has destroyed your motivation, give yourself permission to feel the strong emotions, binge on Netflix, but also to let go. You don't need to be "back to normal" the day after a major loss, but you can take small steps toward normalcy to flex your control muscle: going for a walk instead of a full-blown workout, meeting friends for coffee, spending an hour a day diving back into your work project. Too often, we get stuck in the rut of apathy, because we haven't flexed our hopeful muscle. Small actions that remind you that you have a choice go a long way to training the ability to put your brain back online.

Other research confirms Maier's findings on the role of the prefrontal cortex. Activation of the prefrontal cortex when experiencing pain reduces our negative emotional response. On the other hand, individuals suffering from depression have a reduced ability to activate their prefrontal cortex. Other research shows that Alzheimer's patients who have reached the stage where apathy is prevalent have reduced activity in their prefrontal cortex. All signs point to the importance of this brain area in helping to regulate not only our emotions but also the behavioral apathy that comes with them.

When we have control and can actively choose, we turn on our prefrontal cortex, giving us the ability to regulate our emotional response to stress or adversity. When we lack the ability to choose, our prefrontal cortex learns to shut off, to let the stress response run wild. It's no wonder that we transform into passive responders to whatever challenge we face. We give up. When we take choice away, our brain learns to be helpless instead of hopeful.

In a review on the subject for *Trends in Cognitive Sciences*, researchers Lauren Leotti, Sheena Iyengar, and Kevin Ochsner hammer home the importance of the need for control. In their conclusion, they state, "The evidence suggests the desire to exercise control, and thus, the desire to make choices, is paramount for survival." They go on to summarize their findings: "The desire for control is not something we acquire through learning, but rather, is innate, and thus likely biologically motivated. We are born to choose." We have a basic underlying need to have some semblance of control over whatever we're tackling. Constraining and controlling workplaces take that away, nudging us toward quitting at the first sign of discomfort. Autonomy is the switch that allows us to persist.

Training Learned Hopefulness

"Our goal is to be able to walk by every classroom in the hallway, and each is indistinguishable. You should all be teaching the same subject, using the same activity, at the same time period." These instructions might seem like they are from the early 1900s, perhaps for Henry Ford's innovative assembly line. But as you might have guessed, they came from a modern American school. When I spoke to the teachers, they relayed story after story about how they weren't actually allowed to teach. It was more akin to the paint-by-numbers instructions they gave their students. Their day was scheduled to the minute. Their curriculum provided a script to repeat to the children, verbatim, before giving out the preplanned activity that went along with it.

This may come as a surprise, as schools are filled with mission statements and values of individualized learning and meeting students where they are. But with a rising pressure to improve American students' test scores, there has come a backlash of administrative control. Teachers are thought of as messengers for whatever standardized curriculum the school district spent hundreds of thousands of dollars on. Believing that the magic is in the material, not in the teacher. Efficiency and control have replaced ingenuity and empowerment. As one teacher relayed to me, "I feel like I could be replaced by anyone with a brain. Follow the script. No deviations from the plan. What did I go to school for? My knowledge and experience are wasted." We are taught mainly to follow. Even those who are supposed to be leading and guiding.

In athletic pursuits, it's much the same. Venture down to the local high school or college football field and watch any of those teams train during their conditioning time. The coach dictates precisely what's to be done. The athlete has little to no control. The next exercise to do, the amount of weight to lift, the number of wind sprints to run, the amount of recovery, or even when to grab a water bottle, is dictated and controlled. The athlete is taught to follow orders. For less progressive coaches, there might even be punishment for failing to follow through. If an athlete doesn't complete the last sprint in the required time, he, or the entire team, might have to do another one. Regardless of whether you are on a team with (or observing) a progressive or regressive coach, what's missing from these common scenarios? Choice. The athletes have none.

When it comes to traditional toughness training, coaches envision driving athletes to the brink of exhaustion until they throw up and can't push any longer. This style of training, they often believe, will inoculate players to pain and fatigue. In the workplace, there might not be pushing until puking, but "toughness" is developed through the grind. Junior-level employees work eighty-hour weeks, in the name of proving their worth. As one of my coaching clients, an investment banker, told me, interns and new employees use email scheduling apps to send a few very late-night emails to make it appear as if they were working at nearly all hours. While the *New York Times* reported in 2015 that "emails arrive past midnight, followed by text messages asking why they were not answered." The work culture of proving ourselves, impressing upon our superiors that we, too, can "grind," is prevalent across fields. The employees who survive are the ones who are hardy, who can handle the rigors of the workplace.

But what's missing is that in such scenarios, the players and workers only have one choice, persist or quit. And because of the power dynamics, that choice is often artificial. It's survival for as long as you can. Like the runners on the treadmill for the VO_2 max test, they've limited their ability to be in control. The individuals can't escape, especially if failure, being fired, or being berated is their reward if they don't persist. Instead of training their prefrontal cortexes to be able to turn on, to deal with the negative emotions that fatigue brings about, we are teaching them to be helpless. We've trained them to be the apathetic dog, incapable of hopping over a small gate to safety.

Worse yet, it's not just helplessness. By utilizing such tactics, we're teaching people to respond to fear. If the only reason that you persist is to avoid being yelled at, performing a physical punishment, or getting fired, then the message that's ingrained is to be motivated only when someone is in your face yelling or punishment is on the line. When it comes time for game day, when it's just them, alone out there on the field, what reaction do we expect? The one we ingrained.

Jim Denison is a former athlete and sport sociologist who took a slightly different path in looking at performance. Despite witnessing some of the top athletes on the planet, when he became involved in coaches' education as the director of the Canadian Athletics Coaching Center, he didn't believe that coaches needed more training on how to make someone stronger, faster, or fitter. Instead, he and his colleague Joseph Mills looked for inspiration from Michel Foucault, a French philosopher who had likely never discussed athletic performance in his life.

When I talked to Denison about how Foucault could impact athletics coaching, he pointed to the philosopher's views on power. In Foucault's conceptualization, power was utilized to regulate time, space, and effort. In the political environment, he was critical of the effects of power on the individual. He believed that power influenced control, and when power was relinquished, the individual became passive and docile. Denison and Mills believed the same idea applied to athletes and coaches. As they stated, "Coaching can easily become . . . a technocratic procedure invested less with a coach's understanding of all that training does, and more with his or her power to control, monitor, intervene, regulate, differentiate and correct his or her athletes. Yet, paradoxically, there is nothing structured and certain about a race. What happens over the course of a distance running race is open to constant change; athletes have to make untold decisions that relate to their many bodily states." Denison and Mills continued, "What concerns us, therefore, is how effective a training plan can ever be if the body in the stands—the coach—not the body on the track—the athlete—is in total control of the training process?"

When athletes compete, they are alone in the competition arena. They make the decisions. Yet, in training, the coach takes on the decision-making mantle. Denison and Mills suggested flipping the concept on its head. No, not by making the athletes in charge of designing their workouts, but by giving a large portion of control back to them. By putting the athletes in a position to choose—whether to speed up, slow down, lift another rep, or call it a day—we can take advantage of the power of choice. When we put people in a position to choose, we can "switch on" and train their prefrontal cortex, allowing them to understand and regulate the sensations of pain, fatigue, and anxiety that often come with such difficult moments. We allow them to try, adjust, perhaps even fail, but above all, learn. In Denison and Mills's model, the coach shifts from dictating to putting athletes in a situation where they are challenged, but then giving them free rein to find, search, and choose how to cope with the scenario.

Denison and Mills's ideas have merit beyond better performance. A study of over two hundred men and women found that when athletes trained in an autonomy-supportive environment, there was a correlation with the satisfaction of their basic psychological needs for well-being. Controlling environments were associated with thwarting an individual's basic needs and with lower overall satisfaction. Furthermore, they found that those in a supportive environment tended to have higher levels of mental toughness and better performances.

The key to improving mental toughness does not lie in constraining and controlling individuals. It doesn't lie in developing harsh punishments to teach them a lesson. It doesn't lie in screaming at a person to complete whatever demanding task is in front of them. When we don't have control, we lose the capacity to cope. It's when we have a choice that toughness is trained. Our brain literally turns on, figuring out how to work our way through the situation at hand. We were born to choose, so let us learn how to do it.

TOUGHNESS MAXIM

When we don't have control, we lose the capacity to cope.
It's when we have a choice that toughness is trained.

Training to Have Control

Whether on the athletic fields or in the workplace, if your goal is to train toughness, you have to give people a degree of autonomy. Suffering or experiencing discomfort on its own isn't enough. To train toughness, we need to develop and maintain a sense of control.

Leading Yourself

In the four exercises that follow, I've outlined how to develop that sense of control in yourself.

1. From Small to Large

Take a difficult situation that brings about discomfort; maybe it's performance-related anxiety or a conversation that you are dreading. What we often do is try to control "the thing." So if anxiety is the issue, we try to attack the nerves and fear. We try to force ourselves to turn down the tension, and then after that doesn't work, our brain comes to the logical conclusion that we have no control over our body or the situation. "I can't control my anxiety, so why try?"

Instead of wrestling the giant monster, start with the smallest item that you can have control over that's related to the problem. Is it your breath? Can you intentionally slow your breathing down? Or maybe it's something as simple as showing up on time or getting through the first mile of your marathon. Break it down to something manageable and feasible. The goal isn't to stop there but to get a foothold so that you can gradually climb to the next level. Once you have a sense of control over the smallest item, then move to something slightly larger. Move from small to large.

2. Give Yourself a Choice

Without knowing it, we often box ourselves into a corner, taking any semblance of choice away. We feel trapped, pushed to persist no matter what. Whenever we don't have a choice, we aren't training toughness. If, for example, you say that I have to complete this task by 3 p.m., that might work when completing a manageable task. But when you face something beyond your reach, you're more likely to throw in the towel and say "This is impossible," instead of persisting.

We're often told that when creating a habit, like going to the gym, we need to be rigid and specific. Show up every day at 7 a.m. to work out, and never miss a day. But what research shows is having choice, such as "I'm allowed to miss two days per week if I have to," results in a longer-lasting, more sustainable habit. Katy Milkman, behavioral scientist and author of *How to Change*, calls this allowing for a mulligan. It's why research shows in dieting that cheat days actually help. All or nothing often leaves you with nothing.

Giving yourself a choice sometimes means entertaining the idea of quitting, slowing down, or even giving up. It's not that I want you to do so, but by having a choice, by needing to make a decision, you are developing a sense of control. By considering quitting as an option, you now have influence over the outcome, even if one result is negative. By actively considering quitting, instead of seeing it as something to avoid and never let cross your mind, you are now training toughness. Consider what it would be like to abandon your goal or quit your job.

3. Flip the Script

It was the night before the NCAA regional cross-country championships, and my team was about to face off against the best runners in Texas, Arkansas, and Louisiana. The women's team came in with their best ranking in over a decade, led by a strong trio who were all on the cusp of achieving all-region status. The only problem, senior Meredith Sorensen was in fantastic shape but suffering from one of the worst cases of performance anxiety I'd ever seen. Two weeks before, at the conference championship, while standing on the line awaiting the gun to go off, she turned around, puked all over the ground, then started the race. For Meredith, this was normal. She got so nervous that she couldn't hold her food down. She finished that race in the medical tent with an IV. Two weeks later, she was cleared physically to race and I was stumped about how to help Meredith get in a spot where she could compete up to her potential.

The night before the race, I met Andy Stover at a local bar. Andy was a former collegiate distance runner, and a social worker with a knack for innovative approaches, who also happened to be a groomsman in my wedding. As I relayed all of the traditional techniques I'd tried with Meredith—preparing her for discomfort, visualization, changing her mindset to see anxiety as excitement—he quipped, "Flip the script. Give her back some control." Seeing the puzzled look on my face, Andy continued, "If she throws up before every race, it's become part of the routine. She expects and likely dreads it. So when it happens again and again, all while she's trying everything she can to prevent it, her brain learns she isn't in control. Get her to stop fighting herself. Give her back control."

The next day, as Meredith began her warm-up, she came over and said, "I feel like I'm going to throw up." I replied, "Good! When do you want to do it?" Her face turned from worry to puzzlement. "I don't want to." Seeing her confusion, I replied, "I know, but it's going to happen. So what time would you like to throw up? Should it be before your jog, after you do your drills, or maybe right before your strides? Where would you like to insert throwing up into your warm-up routine? The race starts at 10 a.m. What time should I schedule your puking for?" The confused look on her face was still there, but she seemingly accepted and went along with it. "9:45, right before I do my final strides." Trying to appear as confident in this crazy idea as possible, I replied, "Great, 9:45 it is. I'll set my alarm so we both know and can get it done."

When 9:45 came, my alarm went off, and I walked over to Meredith, telling her it was time to throw up, so let's get it done. The only thing is, she didn't need to. For the first time in several races, no puking occurred. She wasn't perfect or free of anxiety, but she'd wrestled back enough control where her mind was free to focus on what she was doing and not become preoccupied with the anxiety and puking to come. She went on to have the best race of her career, improving by nearly twenty places over her previous best and just missing out on a coveted all-region spot by a mere few seconds in the twenty-minute race.

Away from the running course, you can flip the script by noticing what nudges you toward fear and avoidance. Those triggers are often a signal that we need to flip the script. Feeling like an imposter in your current assignment? Ask good friends if they truly have any idea what they're doing, and chances are they'll tell you they are just making it up as they go. Feeling overwhelmed as you make your company-wide pitch? Tell the audience. At the start of a presentation in front of professional sports coaches and executives, I put a picture of myself as an eight-year-old playing peewee sports up on the screen and said, "This was the peak of my soccer knowledge. Everyone in this room knows more than me, so this is a bit nerve-racking. But I do know about the science of performance, and if you give me a chance, I think it'll help you." When we flip the script, we take away the power of "the thing." We give ourselves permission to do something that we thought was negative. And often, we find this subtle shift provides us with the freedom to perform.

4. Adopt a Ritual

As he stepped into the batter's box, Boston Red Sox star infielder Nomar Garciaparra commenced his painstaking routine. Adjust his batting gloves on both hands, tug on the band on his left forearm, repeat that process again and sometimes again, tap the bill of his batting helmet and various other parts of his body, back to tapping the helmet, make the sign of the cross across his jersey, then windmill his bat until he was ready to swing. Garciaparra isn't the only baseball player with an elaborate batting ritual, or even the only athlete. Tennis stars like Rafael Nadal and Serena Williams have their particular quirks—tying their shoes in the same manner, water bottles placed in the same spot, and so forth, before they go out to dominate. Why do these bastions of athletic superiority resort to painstaking and seemingly silly rituals? Control.

According to the theory of compensatory control, we try to establish order in the outside world in an effort to gain control in the internal. Stepping up to the batter's box, where even the most skilled hitters need a bit of luck, and facing the uncertainty of what ninety-mile-per-hour pitch is coming their way, hitters exert influence over the one aspect of the situation they can control. When we utilize rituals, we shift our focus to behaviors that we are in charge of, pushing to the back of our mind the items that we have little control over. Rituals are a coping mechanism for our brain, convincing it that we have more control than we actually do. If you're completing a task with a high degree of uncertainty and a low degree of control, creating a ritual can be a successful way to keep negative inner voices and emotions at bay.

Leading Others

Much of this book focuses on what we can do to improve our own toughness. But sometimes the degree of control we feel over our life and performance comes partially down to what our boss, coach, or individual in leadership does. If you're leading, you bear the burden of training and empowering control. For that reason, let's quickly cover exercises that will help you develop a sense of control in those you lead or work with.

1. Learn to Let Go

When you're in a leadership position, be it a coach, principal, or executive, it's tempting to micromanage your way to success, detailing exactly when and what everyone should be doing, and where they should be at all times of the workday. Micromanaging occurs because of a fear that someone else isn't going to do the job.

When you dictate and control, you're sending the message "I don't trust you to do the job."

As those in the Special Forces often say, "Trust but verify." It's a balance between trusting and overmanaging, but we often fall too far on the side of overmanaging. After all, our reputation and ego are on the line if we are leading. Instead, let go of the reins a touch, teach them the skills, and then let them go. Check in occasionally to make sure they are headed in the right direction. Over time, the reins should get longer and longer. Your goal is to put people in a position to do their job.

2. Set the Constraints and Let Them Go

Giving away control isn't about letting people run wild with no direction. Set up the boundaries or constraints and then let them go. When a worker is new, this is particularly important. You don't just give someone free rein and hope they come through. Set constraints and let them explore within those, while keeping some control on essential areas that they might not have the requisite skill for yet. For example, in athletics, inform the athlete that they need to complete ten 100-meter repeats, but it's up to them how much rest they have in between. Let them choose. In the workplace, tell those you are managing that you'll work with them to dial in the three most important slides for the quarterly report, but give them the freedom to design and develop the rest.

I always tell anyone I'm working with that my goal is to make myself obsolete. I'm trying to coach them toward independence, not dependence. It might start small, allowing an athlete to decide on one workout every month. But over time, I gradually hand more and more responsibility to those I work with—guiding them along the way, correcting when they fall off course, but never jumping in and declaring, "See! You need me. It's back to me telling you exactly what you have to do."

3. Allow Them to Fail, Reflect, and Improve
Part of giving back control is allowing them to make mistakes. That means giving more autonomy and control in projects that someone can handle. Don't throw them into the deep end with an incredibly demanding task, tell them, "You figure it out," and then watch them sink. Give away control in small bites that eventually grow into something much more significant. Then have a system in place that allows for reflection and growth. In the sports world, after a game, coaches break down the game tape. Good coaches don't scold athletes for making mistakes, as they've already occurred. They use film review as an opportunity to teach someone what to do. In the military, during training, soldiers replace ammunition with paintballs and work through realistic operation simulations. At the conclusion, they go through an AAR, an after-action review, where they break down what went well and what can be improved upon. They don't berate each other for failure in practice. They own it, then get back to work, so they don't make the same mistake again.

Giving people space to fail is something that we occasionally do well in the sporting fields but neglect in the business world. Create conditions that allow for people to mess up and make mistakes in a way that doesn't cause you to lose the big client or big game.

***

When we don't feel in control, our emotions and inner voice spiral. Tasks feel harder, the pain feels more intense, and our doubts seem louder. Our motivation plummets as we head toward apathy and a lack of will to act. It's nearly impossible to be tough when there's no hope of navigating your way through the current situation. It's as if we shrug our shoulders and say, "What's the point?" Being tough is about navigating this experience so that we can keep moving forward. Toughness isn't just about persistence in the face of discomfort. It's about making a *good* decision.

Sometimes the tough decision is to turn around, walk away, and quit. Think of a climber who has spent years preparing for a summit attempt. She gets within a few hundred meters of reaching the top, and her mind and body are urging her forward. "If I can just get to the top . . ." she thinks while she pushes away the fatigue and doubts. The difficult decision is to lean into those thoughts and feelings and recognize that she doesn't have the energy to reach the summit and then make it back down the mountain. The ego is pushing her forward; the tough decision is to turn around. We often equate toughness with persistence, but in some cases, it's the exact opposite. Toughness is navigating the inner turmoil in order to make a good decision. Sometimes that's to persist. Other times it's to quit.

Choice allows us to take back control, to be able to make that decision. It's a kind of superpower that brings back confidence, helps us wrestle with our emotions, and allows us to learn, adapt, and grow. So much so that people report that even if the choice they are given during a task is completely meaningless and inconsequential, that demanding task feels a bit more enjoyable, a bit more manageable. If we want to develop tough individuals, we've got to put them in a position to make decisions, and empower them to do just that. For far too long, we've trained helplessness. The football player who is told to perform burpees until he can't. The teacher who is told to follow the lesson script exactly, and docked when they deviate. We've turned our world, and much of our training, into a version of the VO_2 max treadmill experiment. Survive or quit. It's time to expand instead of constrict. Give people a choice and let them train hopefulness.

THE SECOND PILLAR OF TOUGHNESS

LISTEN TO YOUR BODY

CHAPTER 6

Your Emotions Are Messengers, Not Dictators

Close your eyes and imagine the happiest moment of your life. Perhaps it's the absolute joy of bringing a child into the world. Or maybe it was the day you stood across from your partner, overwhelmed by a transcendent feeling of love and connection as you vowed to cherish and support him or her for the rest of your life. Chances are even recalling that happy moment brings back an assortment of feel-good emotions.

We live for and treasure the moments that bring about love, happiness, and joy. They make life special, worth living. But what if you couldn't experience love or joy? What if you were numb to the sensations or unable to sort through what the butterflies meant while standing in front of the altar?

In 2009, a man named Stephen (not me, to be clear) married the love of his life. Like we would expect, he was smiling widely, a signal of the pure bliss of the moment he was having with his wife. The only problem, Stephen was faking it. Not for a nefarious reason. He truly enjoyed the company of his soon-to-be wife. He simply couldn't feel love or joy. He has alexithymia.

In discussing his condition, Stephen told *Mosaic*, "From an inner-feeling point of view, anything I do that requires an emotional response feels like a fake. Most of my responses are learned responses. In an environment where everyone is being jolly and happy, it feels like I'm lying. Acting. Which I am. So it is a lie."

Alexithymia isn't technically a disorder; it's a catchall term that literally means having no words for emotions. It's intended to capture those who struggle to describe and identify emotions. There's a wide variation of severity in those who experience alexithymia, ranging from individuals who feel nothing to those who experience some sort of sensation but have difficulty putting words to the experience. Stephen echoed this complexity in describing his condition: "I feel something, but I'm unable to distinguish in any real way what that feeling is." Not surprisingly, there are negative consequences for experiencing alexithymia. It's associated with PTSD in soldiers and suicidal thinking in the general population. In the case of alexithymia, it's not just that one's inner body is speaking a different language. It's that the language is in a form we can't even begin to understand. It's like seeing hieroglyphics for the first time without the Rosetta stone. One's inner world is gibberish.

While alexithymia represents the extreme scenario, we all vary in our ability to read, distinguish, and understand our emotions. It's a skill. Our feelings and emotions provide an overview of the homeostatic function of the entire body, a status update of sorts. Emotions help to alert, advise, and regulate. Yet, in the old model of toughness, we're told that emotions should be ignored or suppressed. We shun instead of embrace what we feel. The old model falls short. In order to navigate discomfort, we need to listen to the messages our body is sending.

The Power of What We Feel

Feelings and emotions are close cousins and we often use the two interchangeably. It's easy to get lost in the weeds of the difference between jittery and fear, or unpleasant and disgust. When it comes to toughness, a simple distinction can be made. Feelings are messengers that nudge. They are our body saying, "Hmm . . . something is different," before sending us on a quest to figure out why we feel a certain way.

Emotions, on the other hand, are more complex. They require context and meaning. To go from simple displeasure to sadness, we need to know what sadness entails. We combine a raw sensation (pleasure) with what else is going on in our internal and external world, add in a dash of experience, and all of a sudden, we feel our version of the emotion love. If feelings are meant to inform and nudge, emotions are the alarm bells, screaming at you that something changed and that you need to do something about it. Emotions move us from nudge to shove.

When it comes to toughness, feelings and emotions serve a vital role. Whenever we encounter scenarios where we need to be tough or make difficult decisions, our feelings and emotions set the stage. They bias us toward a particular response. But they don't control us. As writer Robert Wright wrote in the book *Why Buddhism Is True*, "What emotions do—what emotions are for—is to activate and coordinate the modular functions that are, in Darwinian terms, appropriate for the moment." In other words, they are the first step in a cascade designed to prepare us for action.

Imagine you're walking down a dark alley in an unfamiliar city. There's a tinge of unease. The hairs on the back of your neck stand up, and tension rises in your shoulders. You aren't aware of any signs of danger, but your body is on high alert, as if a mugger is hiding around the next dimly lit corner. Where do these sensations come from? Would you experience the same feelings if it were daytime, or if you were walking through an alleyway in your own neighborhood? To understand where feelings and sensations come from, we need to take a peek into how our brain processes sensory data.

Our eyes take in the scenery, noting movement in the shadows, a pothole in the sidewalk. Our nose catches the smells and scents in the air. Our ears pick up the rustling of the grass. Our skin might feel leaves brush against it or the wind traverse across it. Internally, a racing heartbeat, an increase in our body temperature, or even a change in our muscles' acidity provides information that our brain must interpret and understand. Nerve fibers run throughout our body, transmitting information on changes in mechanical, chemical, metabolic, thermal, hypoxic, and hormonal states of just about every muscle, joint, and organ available. In other words, we contain a surveillance system on steroids, like a department store with a sensor attached to every rack of clothing, electric outlet, air conditioner, and piece of equipment in the store. All relaying whenever an item is touched, brushed, moved, or is simply not doing what is expected of it. All feeding back to some master control algorithm that decides whether to alert the security guard, sound the alarm, or let it be.

In the body, this sensory network has a name—interoception—and a location—the brain, stretching from the cerebral cortex to the anterior cingulate cortex (ACC) and prefrontal cortex (PFC) and down into the insula. The interoceptive system provides an overview of the homeostatic function of the entire body, a type of status update on how your body is doing. Is your temperature normal? Are your glucose levels too high? The interoceptive system is active in a wide range of sensations including thirst, touch, itch, sexual arousal, warmth, heartbeat, and even the sensory experience related to wine-tasting. Your interoceptive system helps to alert, advise, and regulate. It's like the instrument panel that an airplane pilot has at her disposal—buzzing, beeping, and displaying. Only, our body doesn't have a digital display, a way to communicate the data of our internal status to our conscious self. Instead, it uses an ingenious method: feelings and sensations.

In 1896, Wilhelm Wundt, the first person to call himself a psychologist, developed the idea of "affective primacy." His theory rested on the idea that small sensations and feelings reached conscious awareness before any other cue. As he stated, "It is the affective elements which as soon as they are strong enough, first become noticeable. They begin to force themselves energetically into the fixation point of consciousness before anything is perceived." Wundt believed that these nearly instant sensations of either pleasure or displeasure guided our actions, pushing us toward either approaching or avoiding whatever triggered the reaction. Wundt's ideas were put on the shelf for nearly one hundred years until modern science caught up.

In his book *The Strange Order of Things*, neuroscientist Antonio Damasio summarizes our current understanding: emotions and feelings "provide us with a moment-to-moment perspective on the state of our health. . . . When we experience a condition that is conducive to the continuation of life, we describe it in positive terms and call it pleasant." Feelings—be they excitement, fatigue, or unease—represent a summary of our interoceptive system, a signal that something is different and that we should pay attention to it.

The function of feelings isn't merely about updating our condition, but to do just what Wundt hypothesized: to drive and direct us toward a possible solution. To approach or avoid something. To eat a piece of fruit (pleasant) or spit it out (bitter/unpleasant). Feelings nudge us toward evaluating whether a signal means danger or we should ignore it and move on. If we listen, our feelings inform and guide us.

Feelings Are Predictive, Not Reactive

Have you ever felt like your phone was vibrating in your pocket, only to swipe at the screen and see no new emails, notifications, or calls? Don't worry. It's not just you; over 70 percent of students had experienced this "phantom vibration syndrome" according to a recent study. Those students experienced a prediction malfunction. Their brain was scanning the environment and created a sensation. We put so much emphasis on the rectangular box in our pocket that we've trained our mind to be on constant high alert, awaiting our next buzz or beep. It's not a coincidence that the more dependent you are on your phone, the more you experience phantom vibrations.

Feelings aren't merely reactive, informing us of what's going on, but they are also predictive, informing us of what's to come. According to the latest scientific theory, our brain predicts the feedback that it will receive. In our alarm-in-a-store analogy, think of it not only as having a sensor watching for feedback, but as someone who guesses which room or glass case is most likely to be in danger at any time. Sometimes, it even flips the alarm before anything has been stolen, anticipating a thief's move. Our brain works in a predictive manner, anticipating what sensory feedback will occur well before our muscles, heart, or skin sends the actual information back to us.

Here's why listening to our emotions is essential to true toughness: they are telling us important information. Our feelings and emotions aren't merely the fuel gauge in our car, but more like the little indicator that tells you about how many miles you have left to drive before the tank is empty. Our bodies are taking in sensory information and making the best guess on what it should keep us informed about. Researchers theorize that feelings and sensations hint at how taxing something we are about to encounter is going to be on the body. How much gas will be drained from our tanks? Feel anxious while waiting to step on to the stage for a performance? That's our body telling us how far out of our norm we are about to push some of our systems, or put another way, it's an indicator of the resources we will need to call upon shortly—no different than the feeling of unease or tension as we walk down an unfamiliar alleyway. Our brain is making a bet that it's better to be safe than sorry, so feelings that trigger potential danger pop up into our conscious awareness, sounding the alarm that we better be ready to run away, even if it might be wrong. When we ignore our feelings in the name of bulldozing through a hurdle and calling it "being tough," we risk not understanding our needs and even our capabilities. We're losing valuable information that could help us make better decisions. Ignoring what we feel is akin to destroying the indicators on our dashboard. No need to know when oil or gas is running low; we'll just guess.

Imagine standing on a wooden suspension bridge, wide enough for a person to squeeze by. With every step, the bridge sways left and then right. You look down and see the trees and river bottom over two hundred feet below. As you peer down, the bridge wobbles beneath your feet, your heart rate jumps, adrenaline starts flowing through your body, and a feeling of excitement or unease washes over you. As you are making your way across the bridge, someone you find attractive approaches. They ask you to fill out a short survey about your experience. As you hand back the completed questionnaire, they write their phone number on the corner of the paper, tear it off, and invite you to give them a call if you want to talk further.

In 1974, psychologists Arthur Aron and Donald Dutton performed this experiment at the Capilano Suspension Bridge in Vancouver, Canada. They weren't assessing fear of heights. Instead, they were evaluating the level of attraction to the man or woman giving the survey.

Upon completing the study, half of the male participants called the number their female surveyor left. When they ran the same experiment on a sturdy bridge, standing only ten feet above the ditch below—a bridge that would trigger fear in just about no one—only 12.5 percent of males made a follow-up call. The discrepancy wasn't caused by any difference in gumption or confidence. It was what researchers call a misattribution of arousal. The men had confused the arousal caused by standing on a shaky bridge with a different kind of arousal, one triggered by the attractive female who had stopped them.

The soaring heart rate and dash of anxiety caused by walking over a high, swaying bridge were (incorrectly) assigned to the beautiful women standing in front of them. Their brain made a quick calculation that said, "Hey! Our adrenaline is going, our heart rate increased, we feel arousal, and there's a woman standing in front of us. We must be attracted to her!" Forget the danger of the bridge. It must be the woman. The men misattributed where the feelings of arousal were coming from. And this phenomenon doesn't just occur when performing acts that might leave us nervous or scared. Research shows that even just exercising (i.e., increased heart rate and nervous system activation) can lead to the same mistake.

Returning to our pilot's flight instrument panel analogy, when it comes to feelings, it's as if our panel doesn't have labels stating what the gauges are for: left fuel tank, right fuel tank, altitude, and temperature. We just have the sensations. We have to fill in their meaning with context. Should we feel afraid or excited? Are we under threat of falling off, or are we having a pretty safe but fun time, like when riding a roller coaster? When it came to Aron and Dutton's experiment, the male subjects got the context all wrong.

They didn't suffer from alexithymia, but it was as if they heard the word literally and interpreted it as being figurative.

Feelings may serve the role of informing or nudging us toward a behavior, but they are also subject to distortion. The better we understand the interoceptive signals reaching our awareness, the better our interpretation—and ultimately the decisions that come from them—will be. In two studies out of Europe, a group of psychologists found that individuals who were clear about their feelings, understanding where they came from and what they meant, were more likely to thrive under stress, anxiety, and pressure. They turned anxiety into excitement and pressure into information and motivation. All thanks to clarity on the message their body was sending.

Feelings serve as our first line of defense. They aren't something that the manliest among us should ignore; they provide vital information. Feelings help us make better decisions. Neuroscientist Antonio Damasio and colleagues set out to investigate the role of emotions in decision making by studying patients who couldn't experience emotions in the same way that you or I would. They found patients who had damage to their ventromedial prefrontal cortex (vmPFC), an area critical for emotional processing. In a series of studies, the patients were put through a barrage of images and stories designed to evoke an emotional reaction. Pictures that generally bring about joy or disgust caused nearly zero reaction. When faced with stories of moral dilemmas, such as whether murder was permissible, the patients with damaged vmPFCs often chose the opposite of the control group. When presented with situations that required evaluating potential harm, they were poor decision makers. They lacked that slight sensation of disgust or displeasure, which guides our decisions not only on what is right and wrong, but in our everyday life. In his book *Descartes' Error*, Damasio outlines how the lab results translated to the real world. The individuals were incredibly poor at making decisions in their day-to-day activities. The patients' lives were often ruined by bad decisions that impacted their familial and work life. When it comes to making decisions, feelings and emotions aren't bad. They are necessary.

The research is clear. The Junction Boys model of toughness that teaches us to ignore or suppress what we feel goes against how our brain and body actually work. The "power through" mantra makes sense only if you actually take stock of what you are powering through. That is what the old definition of grit got wrong. Feelings are signals that need to be understood. Pain isn't something to fear or push our way through; it's a message that needs interpretation. One that sometimes needs heeding, and other times can be allowed to pass by. And if we mistake a challenge for danger or nerves for a full-blown anxiety attack, it doesn't matter how "tough" we are; we are headed straight toward a "freak-out." The first step to toughness is training your body and mind to understand and interpret the signals you are receiving.

TOUGHNESS MAXIM

Feelings are subject to distortion. They depend on context and interpretation. The better we're able to interpret, the better our ultimate decision.

To Feel Is to Decide

By all parameters, the movie *Titanic* was a cultural phenomenon. The James Cameron–directed film netted over two billion dollars at the box office and accumulated fourteen Oscar nominations with eleven victories. Critics and fans alike were blown away by Cameron's masterpiece. But two decades later, controversy still surrounds one decision Cameron made.

In the climactic scene, our heroes, Rose and Jack, are floundering about in the freezing water after the ship went down. They come across a piece of debris, a door, floating in the ocean. Rose and Jack take one quick attempt at both climbing upon the floating door before falling off. After the singular attempt, Jack makes the noble decision to step aside and put Rose's life first, helping her climb aboard and compelling her to stay on while he wades in the water beside her. Jack sacrifices himself, perishing in the freezing water, to save the woman he loves. The ultimate selfless act.

"I think he could have actually fit on that bit of door," Kate Winslet, the actress who played Rose in the film, told Jimmy Kimmel in 2016. Winslet was referring to what has become known as "Doorgate," the rabid Internet-driven debate over whether Jack could have survived by climbing alongside Rose. Internet sleuths have spent countless hours trying to solve this dilemma, mapping out the area of the door and calculating its buoyancy. Even Cameron inserted himself into the fray, stating, "The answer is very simple because it says on page 147 [of the script] that Jack dies."

We can all agree that Jack's actions were noble and possibly heroic. Cameron needed Jack to make the sacrifice, so in the movie world, Jack needed to die. But what if the situation was different? What if, instead of Jack choosing to sacrifice himself, Rose commanded, "No, Jack, get off the door! I need to live!" Would that change how you viewed the scene?

What if, instead of sacrifice, both Rose and Jack repeatedly tried to find a way to float on the door together? But after several attempts, Jack realized that both of them would die if they tried to stay afloat together, so he pushed Rose into the water? How does that make you feel about Jack?

Or, what if our two heroes had been on a door with a stranger, the three of them floating in the freezing sea, but they quickly realized that although three of them were on the door right now, the door simply couldn't stay afloat, so they pushed the stranger off so that they could survive? What if instead of a stranger, it was the film's "villain," the arrogant Cal Hockley, the abusive man Rose escaped?

With each twist of the scenario, you might have noticed a different feeling or sensation. Having Jack and Rose push a bystander off the door to save themselves might bring about a bit of disgust and alter how you see our heroes. When the bystander shifted to the movie's villain, instead of disgust, you might have felt vindication. Despite performing the same selfish action, it now feels justified. When a research team out of the University of Pennsylvania looked at similar scenarios, where violence was either justified (such as the killing of an abusive partner) or not in a movie scene, not only did participants feel different, but their brains also reacted differently. When participants saw justified violence, an area in the brain related to moral evaluation lit up, indicating that the viewer saw the violence as acceptable.

Philosophers and psychologists have utilized such scenarios for decades to evaluate how people grapple with complicated moral decisions. How do we know what is right or wrong? These scenarios are often referred to as "trolley problems," where the ultimate end is the same in all scenarios (i.e., someone dies) but the context on how they get to that point changes. The classic trolley problem involves an individual standing at a train track, with a train barreling down on five individuals. The individual is presented with a choice: he can flip a switch and divert the train to an alternate track with only one person standing on it, or he can do nothing. So instead of killing five individuals, only one dies. In an alternate scenario, instead of flipping a switch to divert the train, our decision maker has to push one individual in front of the train to save the five. Does that change whether or not we would follow through with the action and our views on whether it was moral or not?

In 2001, Harvard psychologist Joshua Greene presented subjects with several variations of the trolley problem while scanning their brains with an fMRI machine. Half of the stories presented contained scenarios where the subject had to harm someone else in order to save the others (i.e., push someone in front of the trolley). The other stories focused on impersonal harm, such as flipping a switch. The former stories tended to elicit a feeling of disgust or displeasure, while the latter did not. When you make it impersonal, the visceral reaction goes down. And when Greene and colleagues looked at the fMRI scans, when subjects read a personal harm story, the areas in the brain related to emotional processing lit up. This flicker of feeling and emotional processing was also tied to the decision each subject made about the scenario's morality. The strength of the feeling predicted the decision.

In his book *The Righteous Mind*, Jonathan Haidt outlines the evolution of moral reasoning since the work of Greene. While we might like to think that logic leads to whether we judge something as moral or not, Haidt believes feelings, not reason, play a vital role. In his book, Haidt concluded, "When we're trying to decide what we think about something, we look inward, at how we're feeling. If I'm feeling good, I must like it, and if I'm feeling anything unpleasant, that must mean I don't like it." And while we've focused on moral judgments, feelings play the same role in just about any decision we make. Feelings like disgust send a message, telling us whether something is good or bad for us. Feelings don't just communicate; they nudge us toward a behavior. Push us toward action. They help us decide what path to choose.

Accompanying any difficult situation is a cacophony of feelings and emotions. Understanding their role as messengers and pushers allows you to find clarity instead of being overwhelmed with confusion. Feelings give you a clue to what your body is predicting. A touch of anxiety signals that you may be a bit wary of what's to come but you're fully equipped to take it on. A sense of dread, and you know your body is preparing for the worst, ready to hit the eject button, and you might need to try a different coping strategy to make your way to the other side of the challenge. Feelings provide clues to where your body is hedging its bet. And the beautiful thing is the more clarity you have, the easier it is to choose whether you go along with that decision or chart a new course.

TOUGHNESS MAXIM

Feelings send a message, conveying information and nudging us toward a behavior.

Appraisal: Reading the Signals

According to a 2015 analysis, nearly 17 percent of teenagers have cut themselves or enacted some form of self-harm. For the most part, people cut themselves not to cause severe, lasting damage but as a way to cope. To deal with the rigors of life, cutting oneself offers a feeling unrelated to whatever issues a person is dealing with. A trading of one sensation (emotional pain) for another (physical pain).

In a series of studies out of Swansea University, psychologists offered an additional explanation for why people self-harm. Instead of focusing on the coping element of self-harm, they investigated the sensations and emotions that trigger the behavior. Hayley Young and colleagues hypothesized that individual differences in interoceptive ability—or the capacity to process and conceptualize the various signals our body sends to our mind—were at play. Their initial study found a connection between those who self-harmed and how they self-rated their interoceptive abilities. The self-harm group had greater interoceptive ambivalence and lower interoceptive appreciation.

Once that link was established, they took it a step further. They had subjects sit quietly and count their heartbeats. The catch: they couldn't put their finger on their pulse or use any other simple solution. This test of our interoceptive abilities forces us to dial in on our most basic internal feedback, our heart beating. The more accurate we are at guessing our heart rate, the better our ability to read the internal status of our body. When Young and colleagues compared those who had a history of self-harm to a control group, the self-harm subjects were more aware of their feelings and sensations, but they performed worse on the interoceptive task. They felt more but couldn't distinguish or interpret what those signals meant. As the authors concluded, self-harm "may serve to resolve the resulting state of emotional and interoceptive uncertainty associated with the body's function in emotional experience."

In some ways, this is a close cousin to the individuals who were fooled by their feelings while standing on a bridge. One was a misattribution of a feeling, the other an inability to process or understand those feelings. If we aren't able to accurately read the signals that our body and mind are putting forward, then to cope with them, we choose the easiest route: ignore or eliminate the sensation. In extreme cases, that might mean self-harm.

Impaired interoceptive awareness has been found in everything from addiction to eating disorders. When we aren't able to make sense of our internal world, we turn to external ways to cope. The same holds true for other feelings and sensations. Kindergartners who don't understand the shame or angst of getting in trouble for the first time resort to tantrums. Or the person who after a frustrating day at work takes their anger out on their partner. When we don't have clarity in our internal world, we tend to resort to less effective coping mechanisms. An ability to read and discern our inner world gives us the flexibility to respond in a more productive manner.

If you've ever gone to the gym after having not worked out in a while, you're well aware of the difficulty in reading interoceptive feedback. Maybe your machismo gets the best of you and harkens you back to your college days, performing a workout that twenty-year-old John or Jill could handle, but certainly not the forty-year-old version. The next day you wake up with pain and stiffness screaming from your muscles and joints. Barely able to lift yourself out of bed and hobble to the door, you wonder whether this is some form of extreme soreness or a catastrophic injury. Do you need to walk it off or head to your orthopedic doctor right away? This soreness-versus-injury conundrum isn't reserved for those of us trying to relive our glory days. It's a vital component of learning for any young athlete. In athletics, we call this knowing what you can train through and what you can't.

An expert at interoception is no different than the veteran pilot who needs to merely glance at a gauge instead of reading the label or manual. An experienced athlete can separate pain and injury. A stage performer can distinguish between nervousness and anxiety. An executive understands when her gut is pushing her in the right direction and when she should ignore it.

Our ability to make sense of the simple (sensations) and the complex (emotions) leads to better decision making and ultimately toughness. Situations that require toughness are those that involve a high level of stress, pressure, or adversity. Such conditions are prime for misreading and misattributing our feelings and emotions. It's easy to mistake a body brimming with adrenaline and excitement for one that's full of anxiety and unease. Research shows that tougher athletes are better able to make sense of whatever feedback their body is giving them. A study out of the University of California San Diego found that individuals who scored lower on resilience had lower interoceptive awareness when put under stress. And in an intriguing study out of the UK, psychologists found that stock traders who had better interoception not only were more profitable but also lasted longer in a business that is notorious for turnover. It wasn't the traders with the better credentials who excelled at making risky decisions; it was the ones who could read their body. When I presented this research to my friend Marcel, who works in a similar field that relies on assessing and making risky decisions, investment banking, he replied, "Pedigree gets you in the door; thoughtfulness and self-awareness are what separate you."

If we continually misread the signals that our body is sending, our brains' predictions on what's coming will also be flawed. Researchers who studied elite athletes and military special forces reported that the "decreased awareness and responsiveness of interoceptive signals leaves [low-resilience] individuals unprepared in the face of interoceptive perturbation. . . . [Low-resilience] individuals may be unable to make accurate body prediction errors, as their reduced interoceptive monitoring may lead to poor integration of current body states to predict future body states." In simpler terms, bad data in means a bad prediction out.

When we lose the capacity to distinguish the nuance of the experience, we jump straight to the easy decision. Part of being tough is fine-tuning your ability to experience and decipher what you're feeling. Better interoceptive skills are correlated with better emotional functioning and linked to lower levels of depression. Those who suffer from depression aren't able to read their body as well as those who don't, much the same way that amateur athletes can't discern between normal pain and a potentially debilitating injury.

Tough individuals develop the ability to discern the nuance that most of us are blind to. Thankfully, this is a skill that can be developed. As we've seen in this chapter, discerning nuance involves two components:

1. Awareness of feelings and sensations
2. Interpretation and contextualization of feelings and emotions

The first step toward developing nuance involves going deep into the experience. If we direct our attention toward an emotion or sensation, examining the feedback for long enough, we can start to distinguish shades of gray where there was once a single category. Go toward the discomfort, deliberately focusing your attention, so that you can peel back the layers.

Another strategy that allows us to understand the nuance of signals and sever the bond they have with actions is to label them. We can interpret very similar sensations (increased heart rate, sweaty palms, a touch of jitters) in drastically different ways—

from nervousness to excitement. When we label emotions or experiences, we can change not only our interpretation of them, but also how our body responds. When researchers out of UCLA had participants label what they were feeling before giving a speech, they dampened down their brain's alarm (amygdala) and turned on the brain's controller (PFC). The more granular their descriptions, the better they were able to handle the swirl of emotions that accompanied the public speaking. Clinical psychologists employ this same concept by utilizing items such as the emotion wheel, a graphic that lists common feelings, such as angry, and pushes us to consider more granular versions, like resentful, indignant, furious, jealous, or withdrawn. How we describe and label what we feel impacts our subsequent performance.

When we name something, we take back control—converting the ambiguous to something tangible that we can understand, manipulate, and come to terms with. Even how we talk about feelings and emotions matters. Take the example of depression. It's common to say, "I'm sad." But that doesn't make sense when you think about it. That implies that sadness is concrete, a trait that you can't change. If instead you say, "I'm experiencing a wave of sadness," it implies that it's a trait that will pass. It might seem trivial, but the language we use to describe what we are experiencing goes a long way in determining whether we have power over our emotions or they control us.

If we can develop the ability to discern what sensations and emotions are telling us, it not only weakens their connection to the cascade of negative thoughts or actions, but it also allows us to interpret them more accurately and to understand that some feelings are meaningless or unimportant—false alarms triggered by an overactive inner self. Something to let float away, that reminds us that a feeling is nothing more than a piece of information. The better we understand the signals coming from our body, the better decisions we can make.

TOUGHNESS MAXIM

Poor interoception → Poor predictions → Lower toughness and worse decision making

---

Developing Nuance Exercises

Exercise 1: Go Deep to Understand Nuance

1. Get specific.
In this exercise, you want to experience feelings that are closely related to the situation you are working on. For example, for the pain of an athletic competition, hop on an exercise bike and do a hard workout or sit in an ice bath and feel a similar kind of pain. For anxiety, if it's social, put yourself in an uncomfortable social situation. Or stand on the balcony at a hotel if heights give you that same feeling. Always make sure you are taking safety into account.

2. Go into the feelings and sensations.
Direct your attention right at what you're experiencing or feeling and sit with that sensation. The goal is to experience without judging. Our aim is to collect experiences. Not necessarily to do anything about them, just to begin the process of unraveling what "pain" or "anxiety" means in different contexts.

Exercise 2: Name It

1. Develop a vocabulary.
When a first grader is asked to describe a person, they stick with basic adjectives: pretty, smart, nice. As they expand their vocabulary, they can describe people and objects more effectively. When it comes to emotions and feelings, most of us act like kindergartners. Expanding our vocabulary allows us to find nuance and clarity.

2. Describe the feeling.
When describing what you're feeling, get creative and try to describe it in as many ways as possible. For example, pain is nebulous. Is it burning, dull, transient, or constant? The same goes for stress or anxiety. Use tools like the emotion wheel, or search for synonyms that may better describe the experience. When describing our inner experience, first go for breadth, then depth.

3. Separate the feeling from the physiology.
While describing, try to separate the physical sensations from the feeling. For example, sweaty palms and a racing heart are the physical sensations. The anxiety or fear that we often intertwine is separate.

4. Name it.
Emotions develop based on context. Find the nuance between the different signals. The nerves you feel before giving a talk? Call that your performance adrenaline. When we name something, we exert power and control over it. We are saying, "I know what you are and how to handle you."

5. **Reappraise it.**
Reframe the signal as helpful information. Can you see anxiety, fear, pain, and sadness as signals conveying a message? Now that you understand the nuance of feelings and emotions and have a name for what you're experiencing, take control of the message. Can you see anxiety as excitement? The fear you experience is a reminder that caution has value. The sadness after a loss is a reminder to cherish and reconnect with those you still have with you. Reframing emotions and feelings as information that you can choose to listen to or simply let float on by is a powerful tool for navigating our messy inner world.

Messenger or Dictator?

When it comes to feelings, we tend to focus on the final step, the regulation of them. Do we ignore them or give in to them? Males are often taught from a young age to ignore or block out what we feel, while females are derided for being "too emotional." Both sexes are sent a similar message, that feelings and emotions are something to be suppressed, except for in particular situations. We are taught to follow our passion when it comes to choosing a job or listen to our heart when it comes to finding love. We receive a contrasting message to listen to some emotions but block out others. The focus is on what to do with them.

Emotional regulation plays a role in decision making and toughness (something we'll cover in chapter 9), but it's the final step in a cascade. By the time we get to regulation, if the bond between what we feel and our subsequent behavioral response is too tight, no amount of willpower or strength can sever that connection.

Like a child who jumps straight from being insulted to throwing a fit, if the bond is too tight, not much can be done.

If we focus on the front end, interpreting and understanding our emotions, we have the potential to impact everything downstream—from where our attention goes, to our inner dialogue, to our behavioral response. We give ourselves a better shot at delaying the jump from feeling stressed to a full-blown freak-out. The ability to read and understand our inner world determines whether we are at a loss, guessing what alarm is going off, or know what message our body is trying to send. If we know the message, choosing the correct solution becomes much easier.

Toughness is about accurately reading these signals—knowing what your body is saying and being able to decide whether or not to respond. It's not that we have to give in to every craving, every signal. Some might be wrong. Others (e.g., the urge to eat sweets) may be a remnant from a past when calories were much harder to come by. Reading your feelings and emotions helps give you the ability to choose whether to give them attention, simply let them pass by, or utilize them for motivation. When testing how individuals work in high-pressure situations, researchers out of Spain found that people could use the anxiety that came along with pressure to their advantage. They could persist longer at a task, reach a higher level of achievement on an academic test, and even have greater job satisfaction. All thanks to the feeling of anxiety. What separated those that were able to use anxiety to their advantage? Whether or not they had clarity on what they were feeling. The researchers concluded, "Individuals who are clear about their feelings are more likely to thrive on anxiety." Even so-called negative feelings can be beneficial. It comes down to clarity of our inner world.

The better we can read and distinguish the internal signals that our body is sending, the better we are able to use feelings and emotions as information to help guide our actions, instead of missing the signal or moving straight from feeling to reacting. A systematic review found that the better our interoceptive abilities, the better we can handle stress. From elite athletes to military personnel to adventurers who experience extreme stress, researchers found that one of the keys to performing under such conditions was an ability to listen to and understand their internal state. They were able to match the feedback their emotions were giving with an appropriate response. Those who seemed to succumb to the perils of the situation struggled to read their internal signals. They were like a new athlete who can't distinguish between pain that will go away and pain that signals a likely injury.

When we have clarity on what we're feeling, we can keep that signal as informational with a little nudge, versus an alarm bell that dictates and enforces. Uncertainty sounds the alarm. Clarity allows us to find the appropriate button to push. Are feelings and emotions messengers relaying information? Or are they dictators pushing us toward a reaction with little or no control over the outcome? Our goal is to keep them largely as messengers.

CHAPTER 7

Own the Voice in Your Head

In 1981, a twenty-one-foot sailboat departed the United States to traverse the Atlantic Ocean. The journey to England was remarkably straightforward. Only a few minor issues arose. They spent some time in the UK before traveling south to Spain and finally to the Canary Islands, located just off the northwestern tip of Africa. In January 1982, it was time to head back. The small boat set sail due west for its return to America.

Seven days after departing, the captain and crewman were jolted by a loud thump in the middle of the night. A collision with a whale, they guessed. The impact ripped a hole in the bottom of the boat, and water rapidly swept in. Dashing into action, they threw the life raft overboard and tossed whatever was within their grasp into it. Before the ship succumbed to the ocean, they made one last attempt to grab anything useful they could find: an emergency kit and as much food as they could carry. By the time the sailboat sank, they were in a life raft that measured six feet across, with a smattering of survival equipment and enough food to last at most two and a half weeks. Adrift in the Atlantic, they came to terms with reality. There were two possibilities for survival: a ship spotting their tiny raft in an enormous ocean, or drifting with the current toward somewhere in the Caribbean—a journey that they estimated would take at minimum two months.

Figuring out how to survive with enough food and water became the main priority. Their rations were meager, so they set about trying to extend them. They fashioned a spear to hunt for fish and a device to purify seawater through evaporation. Even such ingenuity, when combined with the occasional rain's nourishment, yielded about a soda can's worth of water per day, just enough to keep dehydration at bay.

As days turned into weeks, thirst ravaged their bodies and minds. Conversation was sparse, but one day, when the crewman was reaching the brink of despair, a terse exchange commenced:

"Water, Captain. Please? Water . . . more water, Captain. We must have more."

"No! No! Well, maybe. No! You can't have any. Not a drop."

"Please, Captain. Water. Now, before it's too late."

"Okay, the tainted water. You can drink as much of it as you want. But the clear water remains. One pint of it a day. That's the limit."

A stern captain, despite being on the brink of dehydration and starvation, stoically holds true. He knows that if they are to survive, they cannot give in to their immediate desires. They are in it for the long haul with a singular goal: to survive. For seventy-six days, the captain and crewman drifted across the Atlantic Ocean, passing nine ships that failed to spot their small raft or the flare they hastily fired. No one responded to the emergency position-indicating radio beacon. Instead, for weeks on end, the captain would make the tough decisions, holding on to the vision of survival. One pint of water a day, meager rations of food, keeping his crewman focused, and a dash of holding out hope for a miracle. That miracle eventually came, as they somehow drifted across the Atlantic to the tiny Caribbean island of Marie-Galante. They survived, in large part thanks to the ingenuity and steadfastness of the captain.

The only problem with the narrative? There was only one survivor, Steven Callahan. The captain and crewman were the same person. He was performing a solo crossing of the Atlantic. The conversation, though, did take place. As he later reported in his book *Adrift*, Callahan had divided his mind into different characters, a rational, physical self, and an emotional self. The crewman acknowledged the reality of the situation: his pain, fears, and desires. The captain kept him in check and made the difficult decisions that were necessary to survive. As Callahan reported, "My emotional self feels fear and my physical self feels pain. I instinctively rely on my rational self to take command over the fear and pain."

Were Callahan's split characters a result of him going crazy while lost at sea? Was it a survival mechanism in response to extreme peril? Conventional wisdom might think so, but we split into similar characters under nearly any stressful situation. We've all experienced the battle that is waged in our head. The inner devil that complains about fatigue, constantly raising doubts and reasons for you to quit. While another part of you, the angel on your shoulder, counters with motivational mantras and confidence-boosting dialogue to get you to continue to persevere. This inner debate occurs whether we're running a marathon, contemplating whether to buy that expensive dress in the store, or deciding if we should quit our job and pursue our passion. That inner "devil" voice may be trying to pull you away from reading this book and toward something "better" right now. The only difference between our experience of inner dialogue and Callahan's is he gave the two voices a name. While it might seem strange, according to the latest theories in neuroscience and psychology, our mind functions like Callahan's split identities, different selves arguing with each other, competing for attention.

"I can't do this; they'll see right through me" pops into your mind as you sit in the conference room, stomach in knots, waiting for your one shot to impress a major client. Just before you break out into panic, another thought surfaces: "Wait a minute. You're prepared. You know this inside and out. You've got this!" The old model of toughness shuts down this internal debate, preferring to push away or refuse to acknowledge whatever voice we deem as destructive. As if the negative voice is a character flaw, a sign of "weakness" seeping out. But new research shows that both voices are telling you something important. Neither is good or bad; they are conveying information that sometimes we want to listen to, other times we let float by. When we reframe toughness as something defined by awareness of these voices, they become tools that help us make better decisions when things get hard.

The Modular Mind / The Many Selves

We tend to think of our brain as a computer with a central commander sitting at the helm, integrating all the information and making decisions for us. We assume that the different areas of our brain are interconnected, capable of communicating with each other, and that our commander has access to all the data, feedback, and information so that it can make a decision. Unfortunately, this isn't the case. The brain is a patchwork mess.

If we think of how the brain developed, it wasn't an apparatus where each and every part was designed and integrated to function as a whole. Instead, it was pieced together over millennia as humans and the gooey structure sitting in our skull evolved. We added on top of the existing framework, adjusting, assigning new roles, and making it work for whatever demands we faced.

It's akin to buying a historic home built in the 1800s and updating it to meet our modern standard of living. The walls of the house aren't built for central air-conditioning or heating. There were no such things as electrical outlets, cable TV, or even modern plumbing when the home was built. Tearing down the walls and starting from scratch would be unacceptable, so you make do. Trying to find a work-around to give modern conveniences within the confines of the structure you have. Maybe you repurpose components, converting the once-detached kitchen into a garage. You might add in window air-conditioning units or rip up the floor to find a unique way to install heating and cooling systems. Maybe you tear down a bedroom wall, combining adjoining rooms to make a master suite. Or convert the anachronistic "fainting" room popular during the Victorian years into a third bathroom. Regardless of the alterations made, you have to build upon the structure that is already there. You can't design a house that utilizes modern methods of integrating heating, cooling, electricity, and more in the most efficient way. You make do.

The result of a brain that's pieced together is a brain that works as a series of modules—not siloed-off compartments, but instead a mishmash of areas that might directly communicate to one area, while having only indirect contact with another. We're left with a system that can hold contradicting information in various parts of our brain. One module might receive information that our core temperature is rising at an alarming rate, while another module focuses on the strong motivation and degree of importance of our current task. One self pushes us toward completing our goal; the other wants us to stop before we reach imminent danger. If they can't communicate with each other directly, who wins out?

Instead of having a CEO making the ultimate decision, our modules function as a series of subselves, a collection of different areas in the brain that can communicate with each other easily. They are able to work in tandem toward accomplishing different goals. While there are likely numerous subselves, researchers have identified at least seven so far: self-protection, mate attraction, mate retention, affiliation, kin care, social status, and disease avoidance. Whenever we encounter an uncertain or stressful situation, the subself best equipped to handle the situation speaks up.

If you've watched the Disney Pixar movie *Inside Out*, then you are already an expert on the modular mind and how subselves win out. The movie depicts emotions such as Joy, Sadness, Disgust, and Anger as different characters in our protagonist Riley's head. As the emotional characters take in information from Riley's world, they argue among themselves, attempting to wrestle control over the command center filled with buttons that correspond with different actions and behaviors. When one of the emotional characters takes control, he or she presses a command button that pushes Riley to act in a certain way. Just like in real life, our emotional characters have incomplete information. In one scene, Disgust laments, "Riley's acting so weird. Why is she acting so weird? . . . Joy would know what to do." Joy, Sadness, Anger, and Disgust might not always know what the best course of action is, but they argue, debate, and shout, with only one winning and rising to the surface of Riley's consciousness with a push of a button.

As they did in *Inside Out*, feelings and emotions act as the trigger, activating modules, creating internal debate, and pushing us toward an action. When we experience fear, maybe from a wild animal that's approached our campsite, our self-protection module might kick in, engaging the inner voice and screaming at us to abandon our position and run for it. What if instead of being alone at the campsite, we are a parent with a young son? Instead of fear for ourselves, we'd experience fear for our child. Self-protection wouldn't be our primary concern; protection of our child would. Our kin care subself activates, and we feel compelled to stand in the middle of danger, forming a barrier between the animal and our child. We can see the impact of our modular brain in nonthreatening states, as well. In one study, when a group of men watched part of a scary movie, they then perceived pictures of men from different ethnic groups as angrier than if they had watched a relaxing movie. In this instance, the feeling from the movie triggered a defensive and protective module left over from when we had to be wary of those outside of our tribe.

While Pixar took some liberties with *Inside Out*, our current theories of emotions and subselves line up remarkably well with the cartoon depiction. Unlike in the cartoon, though, our emotions don't merely debate to see which one bubbles to the surface of consciousness to win out; a full-blown battle ensues. Joy, Sadness, Anger, and other emotions can be seen as individuals competing for control over our conscious self. In some cases, it's clear when a different subself has hijacked the mind. Think back to an argument you've had with your significant other. Hours earlier, you both were showering each other with love and affection. Now, you're trading barbs and lists of everything that the other has done wrong. Did you transform into two different people? Did you both go crazy? According to Robert Wright, as he outlines in his book *Why Buddhism Is True*, such a situation is a clear "tipoff that the brain is under new management."

Yet, more often, our modular brain doesn't function as a switch that is flipped, where we transform into something seemingly unrecognizable. Our subselves function more like a fight between Muhammad Ali and Joe Frazier—a back-and-forth, with fighters exchanging blows. At some points, it seems as though one has the upper hand before the other fighter rallies back. Our subselves battle in similar ways, attempting to wrestle consciousness away from each other. That's where thoughts come into play.

Our Inner Voice

Have you ever been driving along in your car, and out of nowhere, a strange and concerning thought pops into your head: "What would happen if I turned the car into oncoming traffic?" Or maybe while standing atop a tall bridge or balcony, you thought of what would occur if you jumped. No, you aren't crazy. Over 94 percent of people have similar intrusive thoughts that, if we said them out loud, might get us sent to a psychiatrist. Where do these unwanted thoughts come from?

One theory posits that they are mental simulations—our body evaluating possible scenarios for our current situation, one of those being death. In the aforementioned example, there wasn't a strong emotional driver behind the thought, and you likely had little fear of following through on the gruesome behavior. The thought popped into your awareness and exited, maybe with a tinge of anxiety, which caused you to back off the ledge or remain focused on driving instead of checking your phone. Why did such a strange thought enter your mind if there was little chance of you following through with it?

When it comes to facing stressful situations, our various subselves make simulations on what could or could not happen. According to one theory posited by Wright, modules generate thoughts in our subconscious, and the thoughts that break through and reach our conscious awareness become our inner dialogue. Wright believes that thoughts break through based on the degree of importance. If a particular voice reaches awareness, it's because it has a stronger feeling or sensation behind it. Putting all of this together, if the messenger (feeling) shouts loud enough, a corresponding thought will enter our awareness to motivate us toward a behavioral response or action. The toughness sequence we outlined in chapter 2, moving from feelings to inner thoughts to an urge to act to a decision, should be getting clearer.

Intrusive thoughts are just one part of our vast inner dialogue. Researchers define two main types of inner dialogue: integrated and confrontational. We might experience a singular inner voice reaching conscious awareness, a calm version of self-talk where we list out the tasks we need to accomplish or make a mental note of something we want to tell our spouse. In other cases, we simulate a conversation with a real-life person, working through our talking points and how we expect the other person to respond. These examples are what psychologists refer to as integrated dialogue. In this kind of self-talk, it's less of a debate where there is a winner and a loser and more about working through a scenario—practicing how you might respond, taking different viewpoints into account, and navigating your way through them.

On the other hand, our inner dialogue can appear to be like the prizefight mentioned previously, with competing voices trying to win an argument. Two voices that represent two different selves trying to push us toward competing conclusions or actions. Sitting at a restaurant trying to decide between a juicy burger and a healthy salad, two voices may appear out of nowhere to make their case for why the healthy or indulgent option is best. Psychologists refer to this type of self-talk as confrontational dialogue. A negotiation of sorts occurs, with different voices competing for the "win." In situations that require toughness, confrontational dialogue is the norm. The higher the stakes, the more potential danger, the louder the contrasting selves shout.

We experience these competing voices as individuals having different motives. One might be looking after our health, while the other cares only about the potential reward or pleasure. Psychologist Małgorzata Puchalska-Wasyl, of the University of Lublin in Poland, attempted to sort through our vast array of inner voices based on the emotions and motives attached to them. After analyzing participants' descriptors of their self-talk in a number of scenarios, she narrowed in on five different voices that appeared to be the most prevalent:

- The Faithful Friend—tied to personal strength, relationships, and positive feelings
- The Ambivalent Parent—associated with strength, love, and caring criticism
- The Proud Rival—a voice that appeared distant and success-oriented
- The Calm Optimist—a relaxed voice with a positive outlook
- The Helpless Child—embodying negative emotions and a lack of a sense of control

This isn't intended to be an all-inclusive list. But what it demonstrates is that our various voices tend to serve different purposes. They can be positive or negative, supportive or detrimental, excited or calm, and they can appear to be close to us, or as if they are disconnected. Each voice compels a distinct message, pushing us toward a different behavior. Some inform; others urge. Some are employed to keep us out of danger, others to motivate. Some focus our attention; others try to distract.

It's not just the way in which we describe our inner voices that distinguishes them. Neuroscientists discovered that different types of speech activate different areas of our brain. In one study, researchers found that self-critical dialogue activates a part of our brain linked to error processing and resolution, while self-talk related to reassurance activates areas related to expressing compassion and empathy. In another study, different forms of self-talk were associated with regions related to both speaking and listening. While the neuroscience behind our inner dialogue is still young, it's clear that how we talk to ourselves during stressful situations influences our subsequent behavioral response.

The old model of toughness told us to not even recognize most of our inner voices. If you admitted to thoughts of quitting, or inner doubts on whether you could succeed, you might as well wear a scarlet letter. You were weak. Tough people didn't let negativity enter their mind. Of course, that doesn't jibe with reality. Everyone has an inner devil spurring on fears and doubts. By recognizing what actually goes on inside our heads as we face a challenge, we can prepare for and handle what arises. We can use our inner voice to our advantage.

Our inner dialogue can take on many forms. It can seem like a conversation with a stranger, a command we send to ourself, or a strange but familiar voice that suddenly pops into our head. Our inner dialogue can serve many roles: to motivate, inform, instruct, or push us toward action. In the book *The Voices Within*, Charles Fernyhough explains that our inner voice "can help us to plan what we are about to do and to regulate a course of action once it has started; it can give us a boost in keeping information in mind about what we are supposed to be doing, and in psyching ourselves up for action in the first place." As philosopher Peter Carruthers proposed, our inner speech serves to integrate our variety of systems or selves. To bring concerns and motives to awareness and decide what to do with them.

Our inner voice acts as a safety mechanism, translating our inner world into something we can process and deal with. According to the Hearing Voices Movement, which challenges the notion that hearing voices is a sign of mental illness, our inner dialogue is a way of making what we feel and experience tangible. We may feel stress or anxiety, but we have limited options to deal with the sensation. As Fernyhough argues, "Voices and negative rumination might be unpleasant, but at least they can be engaged with. In which case, the dominance of inner speech might ultimately reflect its evolved role in making the organism resilient to stress."

Voices allow us to do something about what we feel. To actively engage and negotiate. In some cases, this might mean an inner conversation with ourselves, and in other cases, it might turn into a case like that of our marooned sailor, hearing voices that may seem like an entirely different person. In both situations, our body has translated something nebulous like feelings to something that we can actively engage with. We go from being able to only ignore or embrace to being able to negotiate with, create distance and space from, or simply brush off a self not worth listening to.

Whatever form our inner dialogue takes, we have control over how we react to and engage with it. Whether we spiral downward thanks to negative self-talk or brush it off as if it's our "crazy" friend spouting conspiracy-theory nonsense, we can shift our inner voice. We can change the dialogue in productive ways or distance ourselves from the emotional response that might come from the more debilitative voices. We can take deliberate strategies to make sure our inner dialogue is working for us instead of against us.

TOUGHNESS MAXIM

If the messenger (feeling) shouts loud enough, a corresponding thought will enter our awareness to motivate us toward a behavioral response or action. Our inner speech serves to integrate our variety of systems or selves. To bring concerns and motives to awareness and decide what to do with them.

Winning the Inner Debate

What do we do about the voices in our head? Up until now, we've discussed how thoughts arise and why, during the most stressful times, we walk around with an angel and devil on our shoulders. Now, it's time to shift gears into how we manage and utilize our inner dialogue.

"I don't want to be here!" shouts your inner voice as you stand in the corridor, about to be thrust onto the stage. Another voice arises, "You've got this!" prodding you to take those few steps that separate hidden anonymity from center stage in front of hundreds. This adversarial angel-versus-devil act that goes on inside our mind is normal. Sometimes it feels like these voices arise out of nowhere and your only choice is to let them go or combat your inner adversary. Other times, these inner voices are conscious and deliberate, as with positive self-talk and mantras. The key to winning the inner debate lies in utilizing both strategies: handling the voices that seem to arise and using our inner voice that we seem to be able to control.

What does winning the inner debate mean? Sometimes it means listening to the angel on your shoulder instead of the devil. Other times it means letting the negative voice float on by as if it's your "friend" giving a Facebook rant. Remember that thoughts allow us to engage with the internal chaos. Sometimes we want to take up that fight. Other times we want to redirect it. When it comes to winning the inner debate, there are three tactics that we can utilize and develop:

1. Change your voice: inside versus outside
2. Know what voice to listen to: positive or negative
3. Decrease the bond: from me to she

1. Change Your Voice: Inside Versus Outside

Emily was clumsily walking around the room, seemingly oblivious to my existence as I stood in the opposite corner. Her eyes were fixed on an object as she periodically spoke, "Ball . . . get ball . . . toss . . . ball," before completing the action, smiling, and clapping to herself. She wasn't talking to me. She wasn't talking to anybody. She was talking to herself. I was thirteen years old at the time, watching my two-year-old sister wandering around the room, entertaining herself. I had a front-row seat for how inner speech develops.

According to psychologist Lev Vygotsky's theory of cognitive development, we weren't always capable of having these internal conversations. Instead, our inner dialogue develops from our early external speech. My sister's experience of verbalizing what she was attempting to do isn't unusual. It's a natural part of development that every child undertakes. We go through a period of development where our external speech is fragmented and directed, informing us what we are trying to do and occasionally motivating us to do it. Parents and teachers will recognize this type of external dialogue, particularly when a child is engaged with a cognitively demanding task. The child will talk through whatever they are working on, reminding themselves of the next step in their task and what they are trying to accomplish. The speech isn't directed at anyone; it serves to inform, remind, motivate, and push us toward action.

Vygotsky theorized that as we develop, this style of external speech slowly shifts to internal. Initially, he noted the similarities between the two, noting how children used abbreviations and fragments, just as our adult inner voice tends to do. In addition, there's a dialogue quality to both, a sort of internal conversation that takes place. And if his theory is correct, then our inner voice should serve the same role as our external voice does for a toddler: self-regulation and direction toward action. While there's much to be learned, modern research has largely validated Vygotsky's theory of cognitive development, which he first proposed nearly one hundred years ago.

But what happens if we revert to our childlike approach, if we take the internal and make it external? Watch any skill-based sport, and you're bound to see a player talking to himself. A tennis player excoriates himself after mishitting a shot into the net, while a golfer mumbles last-minute points of focus as she lines up her swing. Sometimes this external self-talk is instructional, other times motivational, but it serves the same function that our internal self-talk does. But does it work?

When I was competing at the NCAA cross-country regional championships, I knew I was in trouble. I had a good shot at qualifying for nationals as an individual, and so did a few of my teammates. We had an outside chance as a team if we all put together our best races. Only one mile into the 6.2-mile race, I felt my individual and team chances slipping away. The race had gone out much faster than we had anticipated, and as a result, our team race plan had gone to hell. I was supposed to be next to my teammate Marcel, as we'd finished within mere seconds of each other in every race we'd run that year. Yet, as I glanced up, I saw him in the distance, going after the lead pack, and here I was trying to hold on. Inside my mind, I kept repeating that I was okay, that it was early in the race, and that I didn't need to panic, but I could feel the worry bubble over. This was a make-or-break moment in the race. And all of a sudden, I spoke out loud, "You're good. Look, you can talk. You're not even breathing that hard."

I was caught by surprise that I could say a sentence or two in the middle of a very demanding cross-country race. It was as if my body took the reins off. I felt myself relax and began working through the pack of runners I was with and setting my sights on the lead group that Marcel was a part of. It took me nearly three miles, but I eventually made my way to the lead group of six. As I latched on, I pulled beside Marcel and blurted out, "Don't worry, man, I made it up here." The two Arkansas runners leading the race looked back in surprise. And I got another jolt of energy, as if my mind said, "Hey! You can still talk. You must not be hurting that bad after all!" I proceeded to finish the race in fifth place, five seconds behind Marcel. We just missed qualifying as a team, but Marcel, our number-three runner Scott, and I all punched individual tickets to nationals. And I'd discovered a new tactic to quiet the negative voice that came along with the pain and discomfort of racing.

Faced with fatigue, discomfort, and the pressure of obtaining only one of four spots to qualify for nationals, I didn't resort to powering through. I *processed* through. True toughness is about navigating. It's paying attention to the voices in my head, and making adjustments to address or overcome them. Not blindly pushing through them, but taking the time to see what works in that moment. Sometimes that meant giving more power to that voice, by talking out loud. Other times, it meant letting that inner thought float on by.

Research seems to validate my experience. A group of scientists found that coping statements were more effective when they were verbalized. One explanation for this is that inner talk is cognitively more sophisticated. As we just discussed, it came later in our cognitive development, so reverting to a simpler form of dialogue can ease the burden and deliver a more succinct and actionable message. Like a two-year-old telling himself how to climb the stairs or shoot and retrieve a ball, we are stepping back in time, accessing a deeply ingrained system. Another reason that using external self-talk might work well is that it holds you accountable. Research from clinical psychologist Steven Hayes and colleagues showed that when people used overt self-talk, it made them accountable to whoever is in earshot, as opposed to inner dialogue, which only sets the standard for yourself.

This doesn't mean that we should walk around saying all of our inner thoughts aloud, but occasionally giving yourself an overt pep talk or instructions might be a way to reach a stubborn you that hasn't been paying attention to your inner voice.

2. Know What Voice to Listen To: Positive or Negative

Standing atop a diving board, thirty-three feet in the air, is a nerve-racking experience. For a competitive diver, knowing that you have a few seconds to execute a precise combination of twists and turns makes it even more so. Add in pressure, like if the event is a qualifier for the Pan American Games, and you can imagine the thoughts that might go through a diver's head as she makes her way up the stairs to the top of the platform. Psychologists Pamela Highlen and Bonnie Bennett had front-row seats for the inner workings of forty-four elite divers as they took on this task. Measuring anxiety and self-talk, the scientists discovered a difference between the divers who ended up successfully qualifying for the Pan Am Games and those who disappointingly missed out. The non-qualifiers used more positive self-talk.

We often assume that the way to better inner performance is through positivity. If we can overload our inner dialogue with words of affirmation and positive self-talk, then we'll perform to the best of our ability. Crowd out the doubts and negativity with positive thoughts, and the negative has no space or room to grow. "I've got this!" or "I've worked so hard for this" are common refrains to combat the self-doubt that pops up before a challenging endeavor. There is merit to that argument, as several studies show the performance benefits of just that. But it's not that simple.

In a study looking at positive self-talk, researchers out of the

University of Waterloo found that positive self-talk worked as long as the subject had high self-esteem. If they had low self-esteem, positive self-talk could be detrimental. In other words, your brain isn't going to be fooled by false bravado. We need a degree of belief that what we are saying is true. When it comes to self-talk, if you fake it, you don't make it.

When sports psychologist Judy Van Raalte and colleagues at Springfield College investigated positive and negative self-talk during a number of tennis matches, they found that winners and losers didn't differ in the amount of positive self-talk they used. However, match winners utilized less negative self-talk than their less successful peers. When they dug further into the data, they found that it wasn't so much whether someone had positive or negative self-talk but how they interpreted it. Those who believed in self-talk's effectiveness lost fewer points than those who saw self-talk as largely irrelevant.

3. Decrease the Bond: From Me to She

How challenging is it to keep a six-year-old on task? "It's difficult. They can't focus more than a few minutes at a time, so we are taking brain breaks pretty often," answered my wife, Hillary, who also happens to be a first-grade teacher. And what about if there's a distraction, say an iPad with games, nearby? "Forget it. It's herding cats."

In 2016, in a collaborative study between researchers at the University of Pennsylvania and the University of Michigan, Rachel White and colleagues took on the challenge of testing the perseverance of 180 children who were four to six years old. The kids were given what the researchers told them was an essential task to complete, one that they needed to work hard at to be a "good helper." It also happened to be incredibly dull: press one button if they saw cheese on the screen, and don't press anything if they saw a cat.

The researchers also left an iPad on the table, with some fun games loaded onto it, in case the kids needed a quick break.

Before leaving the room, the researchers gave each child some coaching on how to persevere. They told one-third of the kids that they should think about their thoughts and feelings and ask, "Am I working hard?" The second third were given the same instructions but instead of saying "I," they were told to use their name, such as, "Jill is working hard!" And the final group was told to refer to themselves as someone else they looked up to, for example, "Is Batman working hard?" With the instructions clear, the kids were left alone for ten minutes to work, distract, or do whatever they pleased. The six-year-olds who thought in first person, using "I" to reflect on their work, stayed on task only about 35 percent of the time, choosing the iPad for the majority of their ten minutes. The kids who referred to themselves by their name fared a little better, spending around 45 percent of their time on task. But it was the final group, which focused on Bob the Builder, Batman, or Dora the Explorer as the example of someone who worked hard, who stayed on task nearly 60 percent of the time. The more the child was distanced from his inner self, the longer he or she persisted.

"It's easier to give advice to a friend than to yourself" is an adage that most of us have heard, and it largely holds. Should we quit a job or end a relationship? We're often too close to the issue to have any sort of objectivity. We wrestle over the decision, with our inner voice offering a mix of justifications and rationalizations. Yet, if we see the same situation with a friend or acquaintance, the answer comes nearly instantly. We tell our friend that she needs to drop that guy without hesitation. This phenomenon doesn't just hold true with giving advice, but also in helping us persist and navigate internal discomfort. It can be easily influenced simply by changing our grammar.

The six-year-old children were creating what's called psychological distance. When we use first-person pronouns as part of our inner dialogue, the bond between ourselves and the situation is too tight. When we use third-person pronouns, our first name, or examples of others, it creates space between our sense of self and the situation. We transform into that friend giving advice, not blinded by our connection to the issue. According to work done by researchers from the University of Michigan, first-person pronouns tend to create a self-immersed world, while using words and phrases that create space produces a self-distanced perspective. When we are self-immersed, we amplify the emotional aspects of the situation. Our world narrows, and we get drawn into the emotionality of the experience, setting ourselves up for the negative cascade toward choosing the "easy path" in our toughness paradigm. And according to recent research, a self-immersed perspective causes us to see the situation as a threat. We get locked in on any details that might trigger danger. When we adopt a self-distanced perspective, our view of the world broadens. We can let go of the emotionality, seeing it for what it is, instead of letting it spiral. We see our current predicament as a challenge.

Psychologists have used the same paradigm of self-immersed (i.e., "I can do this!") versus self-distanced (i.e., "Jim/He/You can do this!") self-talk in a variety of stressful situations with adults. From trying to impress a love interest, to giving a public speech, to handling the anxiety over an Ebola outbreak, the results held. When put through stressful situations, if we use self-distanced inner dialogue, it not only helps decrease anxiety, shame, and rumination, but also leads to better overall performance. Our public speaking is judged as better by experts, we are better at making fact-based decisions, we persist on tasks for much longer, and we even have higher levels of wisdom. It can also help with processing past traumatic events.

All from switching from *I* to *you*.

In a study out of the University of Michigan, psychologist Ethan Kross found that using distanced self-talk led to lower levels of emotional reactivity when recalling distressing memories of being abused, angered, attacked, betrayed, degraded, embarrassed, frustrated, rejected, or abandoned. Not only were they reporting lower levels of emotional reactivity, in a follow-up study published in *Nature*, the research group found a lower level of activity in a brain area related to self-referential processing when subjects used third-person self-talk when recalling negative memories.

Using second or third person creates distance between the experience and our emotional response. This linguistic trick allows us to zoom out. When we can create space and broaden our worldview, we slow the path from emotional reaction to inner battle to action. By creating space with a simple change in our vocabulary, we regain control instead of defaulting toward the easy decision.

***

How we interpret our inner dialogue goes a long way in determining its impact. Some people interpret negative self-talk as beneficial. They see it as motivational, poking and prodding them forward—so much so that their inner voice sounds almost like an abusive partner. In my coaching career, I've come across a select few athletes who request that I yell obscenities at them during competitions. They claim the jolt of the harsh language jars them out of their comfort zone. I've also worked with clients where positive self-talk backfires. The jolt of excitement that comes with a realization that "I got this. I can win!" sends their body toward catastrophe. Their brain sees the arousal, forgets that it's "positive," and mistakenly interprets it as a reason to shut down, to not push forward. It's easy to come up with hard-and-fast rules on what we should or shouldn't say to ourselves, but just like with emotions, there aren't good or bad inner voices, just ones we need or don't need to hear at that moment. It's up to us to determine which voice we need and when.

Our inner dialogue is complex. It would be much easier to declare that we should be kind and supportive to ourselves all the time. But as this chapter has shown, we have many different voices, each representing a type of self that occupies our mind. In an essay on emotional care, author and philosopher Alain de Botton suggested, "A good internal voice is rather like (and just as important as) a genuinely decent judge: someone who can separate good from bad but who will always be merciful, fair, accurate in understanding what's going on, and interested in helping us deal with our problems." It's not whether our inner voice is an optimist or a pessimist. It's whether it's fair. If we find our internal negativity holding us back, or our eternally optimistic "You can do it!" voice getting in the way of our seeing reality, we need to broaden our experience.

When we are in the midst of a situation that requires toughness, our goal is to make sure the right self is in charge, that whatever inner voice will push us toward our desired action is winning the inner battle. Sometimes that means we need to combat negativity with positivity. Other times it means we need to tune out our crazy inner self or put distance between ourself and our thoughts. But what's clear is that in difficult moments, how we respond to our inner dialogue is of the utmost importance. During challenging situations, it's all too easy to let the devil on our shoulder win the day and cause us to spiral toward a desire to quit or throw in the towel.

THE THIRD PILLAR OF TOUGHNESS

RESPOND INSTEAD OF REACT

CHAPTER 8

Keep Your Mind Steady

Dan Cleather is a man of contrasts. A deeply thoughtful academic who is as comfortable waxing esoteric and philosophical as he is lifting large, heavy objects. The professor, who also happens to be a strength and conditioning coach, dons a series of abstract tattoos up and down his body. Hidden beneath his shirt is a dragon-like tattoo covering the entirety of his right side. Like most of his colleagues, Dan can lift heavy objects quite well. Look a little closer at his tattoos, though, and you might pick up on the subtle difference between Dan and the "meathead" image of the weight lifter that many possess. The abstract tattoos covering his legs represent movements in tai chi, a practice he holds close to his heart. Dan represents a newer breed of strength coach, a deep thinker with a PhD who is well versed in everything from religion to philosophy. Sitting at a pub in the town of Twickenham, England, we turned our conversation to why he got his tattoos, and he gave a response worthy of his eclectic personality: "Part of it is the meaning. But a part of it is the process. Lying there for hours, feeling the pain, and just having to deal with it. I know that may seem odd, but I swear I'm not a masochist!"

Dealing with pain is deeply intertwined with toughness. While pain is not typically thought of as an emotion like joy or sadness, it functions much the same way. Signals coalesce into a message telling us that something may be off. When pressed on his experiences in getting tattoos, Dan responded, "Sometimes you are lying there for three, four hours. And you aren't sure exactly when it will end. You just have to figure your way through it. Once the artist says he or she is done, you go from being able to last another hour if required to suddenly being flooded with all of these different sensations. You experience it all. It's emotionally exhausting at that point, and if the tattoo artists missed something and needed another fifteen minutes to finish, it would be torture. I couldn't do it." While Dan isn't a monk, he also has an affinity for Buddhist traditions. "The key to being able to handle the pain and uncertainty?" Cleather quipped, "Accepting the pain. Not fighting it."

Monks on a Mission

Antoine Lutz and his colleagues at the Laboratory for Brain Imaging and Behavior at the University of Wisconsin explored the same phenomenon that Cleather experienced: pain. Only the researchers were after the inner workings of the mind, recruiting volunteers to lie in a brain-scanning fMRI machine, all while experiencing pain. Instead of a tattoo, volunteers were subjected to a different kind of discomfort, a hot probe placed on the skin directly below their wrist. While half of the subjects were your average Joe when it came to handling pain, the other half were a bit different. They'd each spent over ten thousand hours performing Buddhist-style meditation.

When the painful probe touched the skin, both the meditators and the control group experienced the same intensity of pain, just above a seven out of ten. However, when researchers surveyed the participants on the unpleasantness (i.e., how much the pain was bothering them), the results were diametrically opposed. The novices rated the same pain intensity as nearly twice as unpleasant.

Both groups felt the same amount of pain, but their reaction to it was entirely different.

Peering into the expert meditators' brains provided an answer as to why. It started before they even felt discomfort. In anticipation of the scalding-hot probe, an area in the brain related to emotional processing called the amygdala lit up in the novices, signaling a threat was on its way. Their monk-like counterparts had a comparatively low response. Before they even felt pain, both groups were preparing in drastically different ways. One was on high alert, readying for catastrophe. The other was aware but decided not to trigger the alarm.

As the painful probe touched the subjects' skin, the experts quickly habituated to the discomfort, decreasing it as they lay in the scanner, while the novices felt their pain grow. It wasn't that the expert meditators were shutting off their response: they had developed a different way to respond. Instead of sounding the alarm, they were taking an alternative route to deal with this foreign sensation. They were actually activating the insula, a part of the brain linked to integrating the significance of the sensations one experiences. Meditation had taught them how to not jump straight from pain to freak-out but to find another path—not by ignoring or forcing, but by accepting and working their way through it.

When the expert meditators were asked about the experience, they didn't respond with tales of pushing through the pain or "toughing" it out. Instead, they described the pain as "softer" with "less dwelling." They had a "greater ability to fully embrace the feeling of pain and . . . let go of the appraisal of what the pain meant to them." The researchers concluded that these individuals had somehow developed the "capacity to flexibly modulate conditioned automatic reactions to an aversive event." In layman's terms, they'd figured out how to turn a nearly automatic reaction into a thoughtful response. They'd reappraised a signal that usually triggers alarm bells to be no different than a mild itch. They were responding, not reacting.

Whenever we face discomfort or adversity, we often jump straight from feeling to freaking out. From the sensation of pain straight to the emotion that often comes with it. True toughness is about expansion instead of constriction. It's fostering the approach of the expert meditators. Not to push against the experience, but to create space between the stimulus and response so that we can better navigate what's going on. It's the child who learns that the frustration from making a mistake doesn't require a tantrum. The husband who can sit with his frustration instead of lashing out at his loved ones. The athlete who can separate the jittery sensations of nervousness from the emotional response of anxiety or dread. How we respond is malleable.

Two main areas in the brain play a role in responding. First, the aforementioned amygdala, which acts as both an alarm system and an interpreter for stressful content, both good and bad. It's not only monks and meditators who show an altered amygdala response. When exposed to painful stimuli, yoga masters were able to turn the emotional aspect way down. For the everyday person, when shown aversive pictures or given a painful stimulus, a lower amygdala reactivity is tied to better emotional control, while depression and anxiety are associated with a hyperactive amygdala.

The counterbalance to the amygdala's panic-button approach to threats is the prefrontal cortex (PFC). While the amygdala might trigger anxiety that wrecks our capacity to execute a task, the PFC acts to regulate emotional responses and maintain our performance on the task at hand. A recent study out of Yale looked at how the brain responded to being under the threat of receiving a shock while playing a predator-prey computer game. Participants showed a strong stress response in anticipation of being shocked, but they maintained performance throughout, thanks in large part to the connectivity between the two brain regions. The brain was able to modulate how it handled the emotional distraction. The PFC (and related brain areas) acts as the stabilizer, stepping in and saying, "We see you feeling anxious, but we don't have to sound the alarm." According to the latest scientific research, the connection between the amygdala and the PFC explains much of the individual differences in emotional regulation.

While monks might have mastered this process, those suffering from burnout live on the other extreme. You are likely one of them. Burnout is epidemic in most Western countries, with surveys indicating that as many as 76 percent of US workers experience its hideous effects. A general lethargy, lack of motivation, and feeling of malaise are hallmark symptoms. Burnout alters how we handle challenges. Put through similar stressful tasks as the aforementioned meditators, those who suffer from burnout have the opposite neural reaction. They have a slightly larger amygdala and a weaker connection to their PFC. With a weaker connection between the alarm (amygdala) and the response system (PFC), they can't step in until it's too late. Burnout trains our brain to react in the exact opposite way that meditation does: a hyperreactive alarm without a "brake" to control the runaway emotional response. The modern workplace is training us to lose control over our inner world.

The same phenomenon partially explains why some performers can rise to the pressure-filled occasion of a championship game and why others seem to lose all ability. As stress and anxiety increase, the PFC shuts down, thanks largely to a flooding of adrenaline and dopamine. Too much arousal and the thoughtful PFC is impaired, leaving room for the instinctive amygdala to take over. If you've ever felt like you've lost the ability to think while experiencing a severe bout of preperformance anxiety, then you know what this inner shift feels like. Clutch performers are able to keep their PFC online, despite rising levels of stress and fatigue.

There's a lot at stake for both the athletes and the burned out. We're all trying to tackle puzzles and problems while improving—and thriving—along the way. Working eighty-hour weeks may seem like the answer, an act of toughness, the necessary work to accomplish our goal. In reality, we're lying to ourselves. Grinding through to burnout, ignoring stress and fatigue, is training our brain to shut down. And with it our ability to thoughtfully navigate the actual challenge. Paying attention and accurately assessing the situation trains our mind to stay steady.

As if prodding with a hot probe wasn't enough, researchers at the University of Wisconsin took it a step further with experienced meditators. They had them confront not one but two stressful scenarios. First, they put them through every public speaker's worst nightmare, the Trier Social Stress Test (TSST), a cruel test developed by psychologists that consists of having people give a speech in front of a panel of judges, whose sole job is to criticize and taunt the poor speaker. The second stressor involved applying capsaicin cream to their body. Capsaicin is the active ingredient in peppers that gives them their heat. By combining a physical and social stressor, the researchers could track their stress response via the hormone cortisol and their inflammatory response via their bodily reaction to capsaicin. At the end of this grueling experiment, the expert meditators had a suppressed cortisol and inflammatory response—a clear demonstration that it wasn't just a change in psychology but biology that explained how they responded to stress.

The research team noticed one more interesting phenomenon.

The meditators weren't tricking themselves or entering some pseudo-real state that allowed them to endure more stress, anxiety, or pain. They weren't using distraction or detaching themselves from the reality of what they were experiencing. The meditators were embracing reality. When researchers compared the physiological data with the self-report data, the meditators had a much closer alignment between perception and reality. The control subjects, on the other hand, had a distorted response. Their emotional response was exaggerated compared to what their physiology predicted. Not only did the meditators have an enhanced ability to regulate emotion, but part of that was due to the fact that they were more accurately assessing the reality of the stress they were experiencing. They had "better accuracy in perceiving their internal state or less emotional elaboration of physiological cues."

The difference between master meditators and you and me doesn't end once the pain is applied. Richard Davidson, psychologist and author of the book *Altered Traits,* has found that the alarm bell in the brain—the amygdala—has a distinctive response to a stressor. There is an initial spike within the first five to eight seconds, and then a decline in activity for the next five seconds as the signal returns toward the baseline. When we think of emotional reactivity, we often concern ourselves with the former, the initial jolt, the sounding of the alarm. But research shows that it's not just whether or not the amygdala is activated, but also how long it takes to return to normal. In a group of over one hundred research subjects, the slower amygdala recovery predicted how they evaluated their environment. Those with a delayed recovery were more likely to evaluate neutral facial expressions as negative. In addition, they were also more likely to display traits of neuroticism.

Psychologists call this phenomenon affective inertia, an inability to let go of a sensation or emotion that has taken hold in our brain. Negative emotions or thoughts compound, triggering an increased reactivity and a more prolonged recovery tail. We can't let go of that snide remark by our colleague, or the harsh criticism of our work the teacher offered in front of the entire class. It lingers, for hours and sometimes days, replaying in our mind. In this book, we've called this experience spiraling, catastrophizing, or the lingering effects of freak-outs. Where our emotions push and pull us in a particular direction. We lose control over our reaction and get trapped in a negativity spiral. As we spiral down, we are no longer responding to the actual stressful event but the reverberating waves thanks to the affective inertia.

When we combine the idea of affective inertia with the research that shows that expert meditators can better coordinate their biological stress response to the reality of the stress they face, a pattern becomes clear. The monks, yoga masters, and meditation experts are responding to reality. They muster the appropriate stress response for a situation and let it do its thing. The rest of us are responding not only to the actual stressor but also to the anticipation and lingering reverberations of it. And to make matters worse, the more we "learn" or hang on to that experience, the more powerful our reaction is the next time we're triggered. As Davidson surmised on *The Ezra Klein Show*, "In some sense, the average person is getting a triple dose of pain (before, during, and after). Whereas the long-term meditator is simply responding when the painful stimulus is delivered."

While the neuroscience is young and ever-changing, the idea that we can weaken the bond between sensation and our response goes back millennia. These core principles make up a large portion of ancient Buddhist practices. From training us to not assign judgment to our thoughts and feelings to teaching us to sit with sensations and experiences, Buddhist meditation practices are focused on helping us handle the inner workings of our complex mind. More recently, authors from Stephen Covey to Viktor Frankl have touted the benefits of creating space between stimulus and response.

Creating space is a tool that we can all learn to use, one that helps us disconnect the initial sensation from the reverberating emotional response. And it's a skill that truly defines grittiness because we are working through a challenge, not blitzing past it. We can change course at every step along the way—feeling, inner debate, freak-out. Creating space is a way to disrupt the pattern, to slow down the jump from feeling to freak-out. We do this through both conscious and unconscious mechanisms that can decrease the alarm or give us more time to work through our inner dialogue so that we don't fall into a catastrophe.

The caveat for much of this research is that most of it was performed on meditators with decades of continual practice. Something that most of us don't have the time or inclination to commit to. Fortunately, research shows that as little as four days of mindfulness training can reduce the feeling of unpleasantness. And mindfulness isn't the only route toward working on this skill. The same skill can be honed when you are working out, watching a scary movie, sitting at your desk, or talking to the barista at your local coffee shop.

Creating space is a concept that's applied in one of the toughest environments in the world—a classroom full of six-year-olds. When my wife, Hillary, started her career as a first-grade teacher, clip charts were the go-to behavior management system. If you misbehave, your clip moves from green to yellow to red. A visual reminder, for the whole class to see, that you messed up. The results weren't promising. "It didn't work. Instead of helping, it made kids feel worse. You took a child and dumped feelings of angst, guilt, and shame on top of an already tense situation. Tantrums soon followed," relayed Hillary.

As the latest science and psychology entered behavioral management, the approach shifted. Now, if a student acts out or isn't following directions, Hillary says, "I first provide them with a choice, asking, 'Can you reset?'" A reset is a momentary pause, an opportunity for the child to think about their behavior or mistake and correct it. Teachers explain and practice resets throughout the year. If the child resets, the teacher quickly moves on. As Hillary summarizes, "Everybody makes mistakes and mistakes are okay. A reset is a chance to think through your emotions and come back online. Children aren't used to or equipped to navigate the barrage of emotions they feel. Give them space to deal with them." And if the behavior continues? "I give them two options. For example, you can start your assignment at your desk or at my table. Or, you can reset now or we can practice resetting at recess together. They feel like they have control as they're picking a choice, but I'm steering their behaviors toward what is acceptable. They can't just say 'No.'"

Since teachers have adopted a modern behavioral approach, children still act out and make mistakes, but they learn, adapt, and grow. Tantrums and fits are down. It turns out that even with six-year-olds, creating space, helping them navigate their emotions, and giving them a choice is crucial for teaching them how to navigate life's challenging moments.

Existential psychologist Rollo May best captured the essence of what we are after when he stated in *The Courage to Create*, "Human freedom involves our capacity to pause between stimulus and response and, in that pause, to choose the one response toward which we wish to throw our weight." With the individuals that I work with, we've even given this search for space a name: creating the ability to have a calm conversation.

TOUGHNESS MAXIM

Respond to reality. For most of us, we are not only responding to the actual stress but the reverberations of it. Tough individuals learn to match perception with reality so that they marshal the appropriate response instead of an exaggerated one.

Having a Calm Conversation

When I was a budding endurance athlete, I had a go-to method for dealing with pain: ignore it until I couldn't anymore, then bulldoze through it. In the early years of my running career, it served me well. I was one of the fastest high school runners in the nation and thought I had this whole running-through-pain thing figured out. But the physical and emotional toll was deep. I'd run myself into the ground, throwing up after nearly every race. I took pride in what I thought was a sign of my toughness—a clear signal that I was pushing myself far beyond what my competitors were. It wasn't sustainable. In races where I didn't have the emotional reserve when it came time to reach down and summon something to get me over the hump, the well was dry. With no other tools at my disposal, I'd watch my competitors glide away, as I'd slow down, at the mercy of fatigue.

As I matured as an athlete and then battled through vocal cord dysfunction, I realized that if I was going to survive, I needed to expand my mental repertoire. Whenever we face an unpleasant sensation or the negative mental self-talk that comes with it, there are four ways to deal with it:

1. Avoid or ignore
2. Fight
3. Accept
4. Reappraise

My initial solution was to combine options one and two: ignore until I couldn't, and then fight through it. This is what I call the bulldoze method, and it's the basis of most of our conceptions of mental toughness. Will your way through discomfort. If you can't, that means you are weak. There's no way forward besides beating your head against the wall until somehow you are magically inured to the experience. The bulldoze method serves as the foundation for old-school toughness. It's why Paul "Bear" Bryant relied on conditioning drills in extreme heat and why coaches yell and scream when your performance begins to falter. Much as my teenage self, many still believe that bulldozing through is the way to go. As we've learned, the latter two methods, accept and reappraise, form the foundation of mindfulness and real toughness.

When we choose to ignore or suppress, we first have to direct our attention to a thought or sensation, signaling to our brain that it must be important. If we actively try to push it away or ignore it, we're doubling down. Our brain doesn't receive the signal that we should move on. It gets the message that something important must be in this signal, so it amplifies it. Anyone who has told either themselves or others to "chill out" or "just forget about it" is keenly aware of this phenomenon. Ignoring, avoiding, and suppressing backfire.

Opening oneself up to experience whatever thought or sensation enters our conscious awareness does not give that sensation power; it drains it of its control. Research shows that when we practice opening ourselves up to discomfort, we are better positioned to handle it. Our brain dampens down the tendency to jump straight from feeling a sensation to sounding the alarm. Acceptance creates space, allowing us to let the sensation float away or evaluate and reframe it.

As my running career evolved and I developed the ability to work toward acceptance, training and racing changed. I no longer had to amp myself up before every competition, hoping to find the will to run until exhaustion. Instead, I had a conversation with exhaustion. While before, the sensation of fatigue and effort was a signal to get ready to fight, now it was feedback. A sign that my body was working hard, that my gas gauge was starting to run low. Previously, I'd experience panic or dread as the pain increased, and I became aware that I might not make it to the finish line at the current speed. Now I could work through it, deciphering what was a signal to pay attention to and what I could let pass by. A sharp pain in my Achilles might mean injury, but a dull burn in my quads just meant I only had one big surge left in my legs. My inner dialogue moved from "Oh shit! This hurts. You're tough. Push through the pain!" to something much more tranquil. "Oh, hey, this is getting uncomfortable. That's okay. It's supposed to hurt. This is normal and expected. Loosen up your arms and stay focused on the task." It wasn't that I'd transformed into a Zen master, immune to pain and anxiety. I still felt the same amount of fatigue, pain, and discomfort. I still had the same inner devil on my shoulder, screaming at me to quit. The difference was I had the skills to keep myself from jumping straight from feeling to freak-out. That tiny lull made all the difference. That's the calm conversation.

A calm conversation is about slowing the world down, creating space between fatigue and freak-out. It's developing the ability to coolly, quietly, and nonjudgmentally work your way through a difficult situation. The calm conversation is a tool used to handle the stress, fatigue, and urge to quit during a performance. Or the anger, fear, and frustration during an intense argument. The more space we can create, the better chance we have to interrupt the downward spiral, to choose the difficult path back up, instead of careening off the cliff.

When using the calm conversation, we won't become immune to the influence of stress and uncertainty. But we can improve our decision making under all conditions. In theory, a calm conversation is simple. When we start to feel a rise in emotions and our mind spiraling toward a freak-out, it's pausing, talking yourself off the ledge, listening to the feedback your internal and external environments are telling you, and slowing the world down. Not by fighting, but by understanding that what you are experiencing is normal, that what you are feeling is telling you something important, and that you have the space to choose how you respond.

The calm conversation creates the space to direct, deflect, or reframe the experience. Everything works in concert. Now that you know what it is, how do we develop the ability to have a calm conversation? It's a two-step process:

1. Create space: spend time alone in your head.
2. Keep your mind steady: develop the ability to respond instead of react.

Step 1: Create Space: Spend Time Alone in Your Head
You find yourself sitting alone in a tiny room, bed and toilet adjacent to one another. There's no entertainment to keep your

mind occupied, or even a window to clue you in to the passage of time. And in some cases, only a sliver of light illuminates the room. You sit alone in the cramped space for nearly twenty-four hours a day, with nothing but your inner thoughts. Where are you?

You've either paid a couple of hundred dollars to participate in a silent dark-room meditation retreat, or you are sitting in solitary confinement. The former is a cure for our mental ills, a "cleanse" of our inner world, promising "to quiet the mind" and giving "the body an incredible rest, supporting its own natural rejuvenation." The latter, a punishment, meant to wear you down, show you who is in control and what isn't acceptable in the prison where you reside. Two experiences, one meant to drive you to enlightenment, the other a barbaric tactic that pushes us to the brink of insanity. One that causes lasting psychological damage, including bouts of isolation panic, post-traumatic stress disorder, and lapses in memory and cognitive functioning. The other, which research shows can improve our perceptual awareness, loosen the bonds of our thoughts and worries to ourself, and lead to positive behavioral change. Both push us to do something we all struggle with: spend time alone in our head.

Even outside of the extremes of isolation, we'd rather be anywhere but inside our head. In a study led by psychologist Timothy Wilson, individuals were placed alone in a room with no phones, friends, or objects to distract them. There was a chair to sit in and a table with a singular item on it, a button. Subjects were informed that if they pushed the button, a painful shock would follow. The choice was simple: either embrace the boredom and spend time thinking, or kill time by shocking yourself and inflicting pain. The logical answer is pretty clear. Mind your business, and be alone in your head for a bit. Simple and easy. The behavior of the subjects told a different story. Sixty-seven percent of men and 25 percent of women chose to inflict pain on themselves rather than contemplate their thoughts for fifteen minutes. One individual pressed the button an astonishing 190 times during the fifteen minutes. That meant shocking himself every 4.7 seconds on average.

The skill of being alone in your head is a foundational piece of developing toughness. And most of us are horrible at it. When we're alone with our thoughts, everything is amplified. The apparent power behind the feelings, thoughts, increase severalfold. Our likelihood of pushing toward rumination and spiraling increases. The solution is pretty straightforward: get used to being alone in your head.

While I am in no way suggesting spending time in solitary confinement, lessons from the extremes help illustrate the adage: the dose makes the poison. A little stress that is within our control and that we are free to escape can push our body and mind to adapt. Even silent meditation retreats can push too far, moving us from acute isolation syndrome to a more dangerous chronic variety if we aren't prepared. Fortunately, to improve our ability to navigate the inner world, we don't need to go to such extremes. We don't need to walk into the weight room and try to squat four hundred pounds in our first attempt. Like those resisting getting shocked, most of us are so poor at being alone with our thoughts that the ten-pound dumbbells will suffice.

Take something that the majority of us do when exercising: listening to music. Exercising is a great time to practice being alone in your head. We've got an array of sensations bombarding our conscious mind for a long period of time. It's an excellent avenue for getting comfortable listening to our inner world. Yet, most of us choose to distract from the inner turmoil.

In high school, Britani Gonzales was an all-star basketball player and a state champion in the 800 meters on the track. When she came to the University of Houston, her talents emerged at the longer distances. When running easy, Britani found that music would sometimes help. It served as a distraction. But as the difficulty of the run increased, something changed. "Once you start hurting, music makes the internal battle harder," she explained to me. "My mind would wander. My pace would drift, and I'd end up running slower. I can't hear or feel the rhythm of my stride or my breathing. My arms are moving back and forth at a different beat than the music. My mind would jump around, from my form to the scenery. Running easy and running when your brain is screaming at you to stop are two different skills."

Music grabs attention and distracts us from other feedback. You might have noticed the same phenomenon when working. Listening to music or maybe even a podcast helps when replying to emails, but it impedes when completing a task requiring deep focus. When we need to expend our cognitive resources on staying engaged, even the subtle beat of the music in the background sends us to sensory overload. When effort is required, you need to be engaged. It's why endurance athletes ditch the headphones to train the capacity to be alone in their heads.

In an increasingly distractible world, we're slowly losing the ability to sit with our thoughts and experiences. When our inner self becomes foreign, we become hyperreactive to anything it says. Our interoceptive awareness declines, as we lose the ability to read and understand our inner world. The explosion in mindfulness books, podcasts, and apps is a consequence of this deteriorating ability. We are looking for a solution to a distracted world, where we increasingly don't need to deal with our inner self. When we train our ability to be alone in our head, to sit with thoughts and sensations, we're better able to disengage from negative stimuli.

To be mindful means to be aware. It doesn't just mean some

form of Buddhist meditation. Britani trained her capacity by spending hours in her head on runs. No music, just her. She wasn't in some meditative state on every run, but over time she naturally developed the ability to shift her focus from her breathing to her form to her inner dialogue to nothing at all. She became comfortable with the sensations, thoughts, and experiences. Clinical psychologists have utilized more extreme measures, such as taking sensory stimuli away in the form of dark flotation tanks, to help reduce anxiety and increase interoception awareness. Similarly, with athletes who struggle with "choking" in sport, I've had clients practice their skill—be it shooting a basketball or putting—in the dark to shift their perception and help them become aware and then navigate their inner world.

We can develop the same skill when we work out, but also during everyday activities like cooking dinner, doing the dishes, or taking the dog for a walk. Notice the sensations and thoughts that arise, trying not to judge or assign meaning to them. Learn to tune in and tune out interoceptive feedback and external stimuli—homing in on your breathing, noticing how your inner dialogue jumps from impatience to jubilation to what you're going to have for dinner.

Scientific research validates such an approach. Cultivating awareness activates an "emotion-regulating network" that includes the amygdala. Researchers out of the University of Wisconsin found that when subjects were confronted with a fearful stimulus, conscious awareness improved emotional regulation and directed subsequent behavioral response by enhancing the interaction between the amygdala and prefrontal cortex, that vital link we've discussed at length. In one study, psychologist Regina Lapate and colleagues found that "awareness seems to 'break' otherwise automatic associations between initial (physiological) reactions and subsequent evaluative behavior." It's not that we always need to be consciously aware or directing our attention. In fact, over time, the process largely takes care of itself. But to develop that capacity, first, we need to train it consciously. To be aware, notice where our mind is going, then direct it. The larger our capacity, the more space we can create between stimulus and response.

Step one to developing the capacity to have a calm conversation is straightforward: spend time alone in your head. Be it on a walk, sitting on the couch, or when standing in line. That doesn't mean control your inner world. Just get used to it. See what it's like to sit with boredom, angst, or whatever other sensation arises. Learn to let your thoughts go where they may before nudging them back to the task at hand. Don't try to do too much. Just be alone in your head. It sounds simple, but it serves as the foundation. We need to develop the ability to be comfortable in our own head. To prevent defaulting toward filling that space by grabbing our smartphone or letting our mind wander uncontrollably. Thanks to hours of repetition, distraction and avoidance are often our default strategies.

Distraction takes little effort, so it's easy to rely on it as a strategy. It's why most of us grab our smartphone at the first moment of being alone. Distract, rather than sit with even a slight feeling of unease while waiting for our friend to show up or the commercial to end. We need to overcome that ingrained reaction and to notice the world around us.

Simply being alone in your head goes a long way. If all you do is leave the phone and headphones at home when you go on a walk, you'll start to stretch your mental muscle. But we can also be intentional about the skills we're trying to develop. I split them into three levels: noticing, turning the dial, and creating and amplifying. This progression builds upon what we learned in developing our interoceptive ability in chapter 6 and our inner voice in chapter 7. Now we're putting the pieces together. Noticing helps you learn how not to jump from stimulus to stimulus. Turning the dial trains you to tune in to and out of the world, directing your attention where it needs to go. And creating and amplifying increases your cognitive capacity to examine and adjust your internal world. These skills serve as your foundation, allowing you to create space at rest before applying them when facing discomfort.

Exercise: Noticing: Practice Boredom
1. Sit quietly in a room with minimal distractions. Sit with eyes open or closed; it doesn't matter.
2. You'll start to feel different sensations, which may then turn into positive or negative thoughts. Sit with them. Don't focus on them or try to push them away. Just see where your mind jumps to and what you have a tendency to latch on to.
3. If you feel an urge to stop or check a phone, experience it. Don't fight it. See if it dissipates or grows stronger over time. Remember, the key isn't resisting. We're just trying to get your mind to experience the sensations and thoughts and not jump to a freak-out.
4. When you start this exercise, aim for five minutes of practicing boredom. As you adapt, gradually lengthen the exercise so that you can sit quietly for fifteen to twenty minutes.

Exercise: Turning the Dial: Increase and Decrease the Volume
1. Perform a task during which you usually utilize distractions. Maybe it's going for a walk or jog, cleaning the dishes, mowing the lawn, or standing in line at the grocery store. Choose an activity during which you usually listen to music or scroll on your phone to distract yourself.
2. Notice where your attention goes. Notice when you feel the urge to grab at your phone or fiddle with something.

3. After you've spent time noticing, try to gently nudge your attention to stay fully on task. If you are washing dishes, try to return your focus to the act of washing dishes. Zoom in to what you are doing, as if you're telling your mind what is essential.
4. Before you stop, practice directing your attention in the opposite way. Zoom out, away from what you are doing. Allow your mind to wander or to be distracted.
5. This practice is designed to create the ability to direct and let go of attention. At this point, it's not that one of the strategies is better or worse. We are trying to develop the capacity to have control over our attention. The hope is that over time, your mind learns to focus on what matters, ignoring what doesn't.

Exercise: Creating and Amplifying Using Imagery
1. Find somewhere comfortable to sit in silence.
2. Close your eyes and visualize yourself doing an activity that brings relief, not stress, such as walking through the woods, going on a hike, or playing a round of golf.
3. Fill in as many details as you can and engage as many senses as possible. See the grass, feel the wind brush against your skin, smell the flowers in the forest.
4. Once again, the key isn't to judge but to experience. This form of mental imagery allows you to hone your sensory and visualization skills. The goal is to increase your cognitive capacity to create, amplify, and adjust your inner world.

Step 2: Keep Your Mind Steady: Develop the Ability to Respond Instead of React

A little over a dozen of the best coaches New Zealand had to offer were gathered in a small building located in Snow Farm, a small mountain escape tucked away from civilization that doubles as a winter-conditions car-testing facility for manufacturers across the globe. For the next four days, New Zealand's brightest, alongside myself and fellow American coach Danny Mackey, would talk athletics. The event was organized by Athletics New Zealand's high-performance staff, and I was expecting the same spiel as the dozens of other conferences I'd attended: talk about training, nutrition, and recovery, and argue over the minutia of conditioning for sport.

I sat upright in my chair opposite a New Zealand coach whom I'd met a day earlier. We faced each other head-on, knees inches apart. Our instructions from Emily Nolan, a top strength coach for New Zealand, were simple. No talking. Sit in your chair. And stare directly into the person's eyes sitting across from you. After a few chuckles, silence commenced, and we attempted to comply. You could feel the level of discomfort in the room. The urge to look away, to crack the tension dominating those first few moments. Nervous movements, eyes darting from staring into their partner's eyes to the forehead to the side, and an occasional giggle were the norms the first minute. People were trying to cope with the challenge they'd been given. Staring at a spot on someone's face or wryly smiling and laughing were ways to ease the burden, to cope with the demands of the activity.

As time went on and people came to the realization that this could last well beyond the minute or two most had initially envisioned, the room's atmosphere shifted. Coping went out the window. Now it was about survival. With no idea how long this could go on, we had to choose a different way to make our way through the discomfort: acceptance. As we passed into minutes three, four, and five, the group's tension dropped, faces and bodies relaxed. We stopped trying to fight the unease of looking into a stranger's eyes. We embraced the discomfort. Coming to peace with it until nearly ten minutes went by and we were allowed to stop. Sighs of relief let out.

By now, you should understand the connection between such an activity and toughness. It serves as a beautiful demonstration of putting your body in an uncomfortable situation and then working your way through it. There's no decision to be made, but if you let the charade carry on for long enough, your mind will stop trying to struggle or cope and accept the circumstances. That doesn't mean you "give in." It means that you learn how to be comfortable with the uncomfortable.

I've run this experiment on athletes and professionals, and the same pattern occurs: cope until they all come to terms with the task and accept it. The first time I had come across this idea wasn't in athletics or sports psychology but in a search for love. In a classic study, psychologist Arthur Aron attempted to create a connection between individuals of the opposite sex by having them answer a series of ever more intimate questions about themselves before ending with four minutes of staring into each other's eyes. Aron's research demonstrated that such a procedure increased "closeness," primarily due to the combination of forcing vulnerability and total awareness. How well did Aron's approach work? The first participants to go through the procedure, who had no clue what they were participating in, fell in love and got married.

Upon reading the study in my twenties, the science nerd that I was tried it out on a first date. Despite my enthusiasm and staring at a stranger for four minutes, the results weren't promising. Still, my own failed attempt at love led to the "Aha!" moment of how awkward and uncomfortable the whole experience felt, especially the eye gazing. I'd found a new exercise for making people feel uncomfortable.

Anything that causes slight discomfort and unease is an opportunity to train your mental muscle to create space. You can utilize physical discomfort (e.g., a wall-sit contest, dunking your hand into ice, or holding your breath), lean into fear (public speaking, fear of heights), or sit with the pull of angst (place your phone facedown on the desk in front of you, no touching it while it beeps and buzzes). The actual activity is less important than the feelings that come with it. The goal is simple: put people in an uncomfortable space, one where their anxiety and stress rise and their mind searches for an out. Then have them utilize the skills we've discussed throughout this book to keep their mind steady, to respond instead of react. Train them first, then put them through situations where they apply what they've learned.

Our brains are pattern-recognition machines, and there can be tight bonds between sensations and emotions, or emotions and inner dialogue, or any of those and a particular action. For deeply ingrained mental habits, your brain might skip right from feeling a particular sensation (e.g., angst) to an action (e.g., grab your phone). To weaken that bond, we need to create space. And once we have space, we can redirect toward a more productive response. That's what the calm conversation is all about.

We need to put ourselves in a situation that elicits negative thoughts, sensations, or emotions, then utilize mental skills to work our way through them, to convince the brain that it's okay and doesn't need to go down the well-worn path of a freak-out. This starts with general situations that might not be tied to your particular challenge but that simply put you at unease. That allows you to work on creating space, being nonreactive, utilizing positive self-talk, or zooming your attention in and out in a safe environment where you have little fear of failure. Hopefully, by now, you can see why this conceptualization of toughness is so different from the one traditionally presented. If we try to rely on avoidance or fighting through, we often strengthen the bonds between feeling, emotions, inner dialogue, and action. The brain says, "Hmm, we are really trying to ignore or take on this thing. It must be important."

Practice Having a Calm Conversation
1. Experience the sensations. Practice nonreactivity to them, interpreting them as information.
2. See where your thoughts try to go. Watch them arise and let the negative ones float away. Try to calmly respond with positive dialogue.
3. Feel the urge to quit or stop. Once you feel the urge to quit, try different strategies to navigate the urge. Sit with it, direct your attention toward or away from it, or utilize self-talk to work your way through it. The key is not fighting it.
4. Your goal is to create space between feeling the sensations and jumping toward the urge to quit. You're trying to decouple the feeling and the response.

Applying the Calm Conversation to Your Situation
After the 2004 Olympics, psychologist Hap Davis took a group of elite-level swimmers and stuck them in an fMRI machine to scan their brains. Similar to a post-competition film review session that you'd see in football or basketball, the swimmers were shown videos of when they failed. Races where they'd fallen short of their goal, missed out on making the Olympic team, or let their team down. When watching their failures, the swimmers' amygdalae lit up, with only a small activation in the brain's motor cortex. Their brains were sounding the alarm, triggering a reaction that

amplified the negative emotions related to their own failure. After noticing the trend, Davis took the athletes and put them through a brief training program designed to rewire their response to their failure by understanding and evaluating the emotions and their response to them. After the intervention, the swimmers were once again subjected to watching their worst performances. This time, the internal response differed, with a smaller amygdala and higher motor cortex response. Davis told *Time* magazine, "Watching the failure washed out the negative emotion. Now I can discuss it with you, and it's no big deal."

Once you understand the calm conversation, start small and general, then move toward larger and specific. If your nemesis is public speaking, begin with practicing the calm conversation in any situation that brings feelings of discomfort or angst. That could include physical discomfort, as in submerging your hand in some freezing water, or mental discomfort, such as watching an embarrassing video of yourself. The goal is to just learn how to sit with, then navigate your inner world. From there, move toward specificity. You could start with the nerves of introducing yourself to a stranger at the local coffee shop before practicing presenting in front of friends and family. The goal is to progressively develop the skill in situations that are closer and closer to what you might face.

Mistake watching is a nice transition from feeling general discomfort to experiencing a situation connected to what we do. You can watch game footage, review a presentation, or even go over your sales report with a friend or colleague. As you experience the swirl of feelings and emotions, sit with them, acknowledge them, label them (as we discussed in chapter 6), reframe and learn to see them as old friends conveying information. Maybe you choose to reframe the inner dialogue or create psychological distance by imagining you're watching someone else. It's all about slowing your inner world down so that you can then go about deciding what's worth listening and responding to and what you should just let pass on by. By practicing the calm conversation while watching your mistakes, you are actively desensitizing your mind to the stigma. It's learning to turn on your kinder, responsive brain and turn off your triggered, reactive one. If you find yourself fighting the discomfort during this practice, step away. Fighting backfires; to turn down the trigger of the alarm, we need to convince our brain that our mind is steady and we don't need to flip the alarm.

Training your mind is just like training your body: gradually progress to more realistic and more difficult situations. Be creative. Whatever puts you in a place where emotions swirl and negative self-talk takes hold is an opportunity to employ the strategies we've discussed. Finally, practice utilizing the calm conversation in areas that are specific to our challenge. For some endeavors, this is easy, while others require creative thinking. The goal here is to mimic the sensations as best we can. If anxiety is the issue, then find something that triggers a similar level of anxiety. The following exercises provide a step-by-step guide for putting yourself in an uncomfortable situation and training your mind to work through it.

Exercise: Mistake Watching
1. Watch a video of yourself performing at whatever it is you do.
2. As you feel embarrassment or frustration in watching yourself fail, take note of the sensation. Evaluate the emotions that come with it.
3. Sit with the feelings and sensations. Try to create space and keep your mind from spiraling. Practice labeling the emotions, using psychological distancing, or any of a number of strategies we've discussed so far. Watch where your thoughts go and gently pull

them back away from the "freak-out" stage. Utilize the strategies you've learned, from breathing exercises to mindfulness training to redirecting your attention.

Exercise: Let Your Mind Go to a Bad Place While Performing
1. The only way to get better at something is to practice it. In this activity, practice whatever task you are trying to improve your toughness on.
2. While performing the task, let your mind spiral. Go toward the negative. Feel yourself about to experience a freak-out. For example, as you rehearse your big presentation, go toward the "devil" in your head that's telling you that you don't know what you're talking about.
3. As you experience yourself spiraling, try to pull yourself out of it. Try to use different strategies, seeing what is effective in the moment. Coping strategies will be explored in the next chapter, but a few to keep in mind:
   i. Zoom in/out: Change your attention, zooming in to where you're focused only on a narrow slice of the situation, or zooming way out so that your view is broad.
   ii. Label: Name what you're feeling or experiencing. Remember back to chapter 6: labeling takes away the power of "the thing." The more nuance and clarity you can give your mind, the better.
   iii. Reframe: Alter how you are viewing the situation. For example, are you seeing stress as negative or positive?
   iv. Adjust your goal: Break down your goal into something manageable. Get to the next mile marker on your run or to the next section of your presentation.
   v. Remind: Go back to your purpose for performing the activity. Remember why you started and why it matters.

vi. Give yourself permission to fail: During any sort of challenge, we often go further into a self-protective mode when we become afraid to fail. Freeing yourself up to fail can free you up to perform to your potential.

Finding Equanimity

Upekkha is a concept that Buddhist monk Bhikkhu Bodhi described as such: "A spiritual virtue, *upekkha* means equanimity in the face of the fluctuations of worldly fortune. It is evenness of mind, unshakeable freedom of mind, a state of inner equipoise that cannot be upset by gain and loss, honor and dishonor, praise and blame, pleasure and pain." The concept is so strong in Buddhism that it's considered both one of the four divine abidings and one of the seven factors of enlightenment. In Buddhism, it outlines how to deal with change, or in our terms, discomfort. *Upekkha* translates to "equanimity," which the *Oxford English Dictionary* defines as "mental calmness, composure, and evenness of temper, especially in a difficult situation."

The Buddhist religion isn't the only one to value equanimity. Hindu scripture proclaims to "perform your duty equipoised, O Arjuna, abandoning all attachment to success or failure. Such equanimity is called yoga." In Stoic philosophy, the concept of ataraxia—a state of serene calmness—is emphasized, while in Christianity, the Bible is littered with references to having patience in order to persist. One example can be found in the book of Luke: "In your patience possess ye your souls." Next to this passage, John Wesley, the founder of the Methodist Church, wrote in his 1765 commentary, "Be calm and serene, masters of yourselves, and superior to all irrational and disquieting passions. By keeping the government of your spirits, you will both avoid much misery, and guard the better against all dangers." In other words, equanimity.

We started this chapter with monks and meditation before meandering through the world of athletics and love. It's only fitting to end back where we started. Creating space, being tough, and responding instead of reacting are all forms of equanimity. The concept of equanimity often gets confused with not experiencing, or even suppressing, emotions. But as we learned, even the deepest of meditators experience love and fear to the same degree. They just respond to it in a different way. It's the ability to keep your mind steady, so that you can choose how to respond. Whether we call it equanimity, a calm conversation, patience, or responding instead of reacting, cultivating the space to choose the way forward is key to developing toughness.

In his book *Transcend*, Scott Barry Kaufman, defined *equanimity* as "a cultivation of mindfulness and observation, of not pursuing one's purpose with blinders on but constantly being open to new information, constantly seeking wisdom and honest awareness of reality, and constantly monitoring your progress and impact on your own personal growth as well as the impact on others. . . . Radiating warmth and openness as you encounter the inevitable stressors of life." Equanimity and toughness work in concert.

CHAPTER 9

Turn the Dial So You Don't Spiral

It's YOUR fault!" "No, you didn't tell me!" "Yes, I did!" "Why can't you do anything right?!" Relationships and arguing come hand in hand. Even the most easygoing and stable among us find ourselves in heated debates with our loved ones. Sometimes, it's over something important—financial troubles or your future plans. But often it's about something with far less meaning. Someone forgets to take the trash out or pick up the flour at the store, anger and frustration mount, and before you know it, World War III erupts in the living room. A seemingly superfluous trigger causes an avalanche of emotions and feelings that fail to dissipate until nearly everything is destroyed.

It's as if we revert back to our early childhood, throwing a tantrum because we didn't get what we wanted or got caught stealing from the cookie jar. And from the outside looking in, the adult version often looks as ludicrous as the tantrum. Ask any teenager who's witnessed their parents blow up over something inconsequential. During such fits, teenagers often roll their eyes, while relaying to their siblings, "Mom and Dad are arguing over the dishwasher. Let them chill out. They're in crazy mode." But to the adults arguing, it seems real, and of dire importance. It may start small, but as the swirl of thoughts, emotions, and feelings builds, even the inconsequential can grow into a giant snowball hurtling down the mountain.

When the avalanche starts, when our brain seems to go off-line, we often feel powerless. Like anger has taken over and we have no option but to wait until it dissipates. When we're at our breaking point, when exhaustion from parenting, work, or just life takes over, the old model of toughness fails. Resisting, pushing through, and playing through the pain are all akin to telling your parents, child, partner, or whoever is in the midst of their tantrum to "calm down." It never works, and often feeds the monster.

There's a better way to deal with this cascade of events. Whether it's arguing with loved ones, feeling so overwhelmed that you don't feel like you can step foot in the conference room, or becoming so frustrated that you are about to quit, when we're at our breaking point, we still have options.

Up until this point, we've covered how to rethink our relationship with toughness. How we can develop the mindset and skills to navigate adversity. This chapter outlines what to do when we're at our breaking point, when the response or reaction is upon us. When burnout, blowup, or freak-out is the next seemingly unavoidable step. When the snowball is rolling down that mountain, picking up steam, and barreling toward destruction. What do we do to stop the potential avalanche—divert it, destroy it, or slow it down? The answer lies in learning how to cope.

Coping with Fear

Surrounded by a group of friends, Moise Joseph and Tom Abbey stood out. They came from different backgrounds. Mo was Haitian; Tom was from upstate New York. But with one look at them, you could tell they were phenomenal athletes. Mo was an Olympian. Tom had fallen just short in an elite development program to become one. Both had dedicated a large portion of their lives to attaining fitness levels that few of us could imagine. They'd competed in front of large crowds, overcome anxiety, stress, and fear to perform at their best. They were both risk-takers. Mo had moved across the country to pursue his dream to become the best athlete he could be. Tom had done the same in the athletic world before taking the entrepreneurial route and founding his own business. But as they entered the old rickety haunted house in rural Virginia, their mindsets couldn't have been more different.

Mo was visibly shaking, a nervous wreck. As the group entered the home, his eyes locked onto the person in front of him, and his towering frame became diminished. The six-foot-tall Olympian tried to shrink behind the five-foot, six-inch women who were also on the tour of the haunted house. When "zombies" popped out of the dark, Mo would dart to the other side of the room, letting out a scream along the way. As he neared the finish and found himself in a pitch-dark room, his impatience grew, and he called out, "Move faster, move faster," to any who would listen. Until, finally, the dam broke. He could take no more. At the buzz of a chain saw, unable to contain his fear any longer, he bolted for the exit, using his world-class speed to fly by any actor who dared step in his way.

Tom, on the other hand, was excited. With every disfigured actor or scary clown that jumped out of the shadows, Tom's intensity climbed. He was fully immersed. Like Mo, he let out screams, but they were of a different variety. They didn't come from a place of fright, but of a strange mix of excitement and deliberately developed fear. His eyes were wide open. Tom wasn't trying to survive or get to the exit; he was absorbed in the activity. Taking the clichéd advice to live in the moment to its logical extreme. Letting every ounce of fake blood and each severed body part take him to the next level. Tom was fully immersed in the experience.

Tom and Mo both experienced the same emotional state: fear. But how they handled it differed entirely. Tom amplified; Mo tried to cope and turn the volume of the experience down. A group of researchers out of Aarhus University in Denmark discovered that haunted house visitors could be divided into two categories: adrenaline junkies and white-knucklers. One amplified; the other suppressed. Upon completing the frightful experience, both groups have remarkably similar levels of satisfaction and enjoyment, but they have markedly different experiences. Adrenaline junkies achieve enjoyment by attempting to maximize their emotional arousal. On the other hand, white-knucklers reach nearly the same satisfaction by employing the opposite approach, minimizing arousal.

One group upregulated their fear; the other dampened. Using a combination of video footage and interviews as participants made their way through a haunted house, the researchers found that each group utilized an assortment of cognitive and behavioral approaches to achieve their goal of adjusting the emotional volume. But if we peer further into the strategies, a commonality arises. How they directed their attention played a large role.

When we are faced with fear or any other form of discomfort, our coping strategy influences our experience and our behavioral approach. We can turn the volume up, diving into the experience, or we can turn the volume down, directing our attention away or reframing it, reminding ourselves that "it isn't real." It's not that either strategy is right or wrong. The question is whether it matches the situation and our goals. Sometimes the answer is to be like Tom, amp up the experience, so we feel the thrill of being frightened. Other times, we need to turn the volume down, to keep our mind from being overwhelmed. Coping strategies allow us to maintain the right balance during ever-increasing levels of stress, keeping our brain and body in the sweet spot to be able to perform.

Sitting atop the top ten strategies used by adrenaline junkies to amplify their fear was to "focus on the situation," where participants stated they "tried to be in it." The rest of the list included tactics like using active immersion, maintaining visual attention, staying engaged in what the actors said and did, and reminding yourself that it wasn't real. On the other hand, white-knucklers took the same strategies and flipped them on their head. They tried to imagine it wasn't real, broke visual attention, resisted immersion, directed their mind to something else, and focused on getting to the finish.

These are coping strategies designed to help us turn up, or down, the dial on sensations, feedback, and even our inner dialogue so that we can navigate our inner and outer worlds. They allow us to deal with, accept, or ignore the experience so that we can survive, thrive, or simply get through the discomfort. Broadly, we call these "emotional regulation" strategies, but I think that definition is too narrow. As we've learned throughout this book, the various sensations, emotions, and thoughts work in concert with one another. To regulate emotions means to influence the rest.

We can be like the thrill-seekers and fully immerse in the experience, reframing the potential fear as excitement, or we can be like the white-knucklers, ignoring and suppressing our inner and outer worlds so that we make it through the experience. What coping strategy we use is up to us. But choosing the right one can keep us from reaching our breaking point, from spiraling out of control, from sprinting through the woods at the sound of a chain saw with no chain.

Attending to Discomfort

When he lined up for the 1972 Olympic marathon, Frank Shorter was chasing history in a race that few understood and most saw as an event reserved for those with a screw loose. Shorter was in search of Olympic glory, attempting to win a race that hadn't seen a champion from the United States since 1908. Marathon running wasn't the mass participation event that it is now. It was in its infancy. The first New York City Marathon took place in 1970 and consisted of 127 scantily clad runners traversing Central Park. Even the historic Boston Marathon, which had begun in 1897, sported only 1,219 finishers in its 1972 edition. Quite a bit smaller than the nearly 40,000 runners that make their way to Boylston Street today.

Two hours and twelve minutes after the gun fired, Shorter would capture the elusive title, winning the Olympic gold at the Munich games. Shorter's victory was the spark that set off the running boom. Within a few years, running would transform from something that a select breed of athletes would do to a mass-participation event. Marathons and other road races sprang up in every city, and major races exploded into a festival-like atmosphere with tens of thousands of participants.

Shorter's success and the explosion of running brought a search for understanding a sport that, until that point, had mostly been left on the fringes. Following the 1972 Olympics, a dream team of scientists assembled to unravel the mystery of the world's best runners. The list of scientists is a who's who of the pioneers in the field of exercise science: David Costill, Peter Cavanagh, Kenneth Cooper, and others. The scientists knew they'd struggle to enlist the best athletes in the world to be poked and prodded, so they brought in *Sports Illustrated* writer Kenny Moore to help with recruitment. Moore just so happened to be the fourth-place finisher in the 1972 Olympic marathon. Moore helped recruit the best runners in America to participate in a barrage of physiological, biomechanical, and psychological tests. Frank Shorter was there, as was the legendary Steve Prefontaine, and a slew of other Olympians. The work that came out of this meeting of the minds would define our understanding of sports science for a generation.

Two researchers, William Morgan and Michael Pollock, were tasked with picking apart the runners' psychology. While the masses were beginning to take to the streets, the cliché of the lonely long-distance runner held firm. Runners were introverted athletes who didn't mind spending hours tolerating ever-increasing amounts of discomfort. Morgan and Pollock wondered what set apart athletes like Shorter, who could master the psychological demands of the marathon. They threw an alphabet soup of questionnaires at the elite runners: the State-Trait Anxiety Inventory, Somatic Perception Questionnaire, Depression Adjective Checklist, Profile of Mood States, Eysenck Personality Inventory, and a slew of others. They had a chance to probe the minds of the world's best, and they took it. They were looking for anything that set apart this collection of elite marathoners from everyone else.

After all of the questionnaires, they finished with an hour-long interview designed to unravel the runners' motivation and experience during races. The last question they asked was: "Describe what you think about during a long-distance run or marathon. What sort of thought processes take place as a run progresses?" The researchers, eminent in their field, hypothesized that these world-class runners must use "disassociation of sensory input"—in other words, tune out, distract themselves, turn their attention to anything but the painful, tedious act of running for over two hours.

Contrary to their hypothesis, the elite runners didn't zone out. They did the opposite: they homed in, using an associative strategy. They concluded that the runners "paid very close attention to bodily input such as feelings and sensations arising in their feet, calves, and thighs, as well as their respiration. . . . Pace was largely governed by reading their bodies." On the other hand, the recreational runners in the study chose to dissociate, seldom paying attention to the actual act of running. Some would recall childhood memories. Others would "write" letters to friends, count, focus on the passing scenery, or sing songs in their head. The classic view of association versus dissociation dominated psychology and performance literature for the next several decades. It didn't apply to just runners or even athletes. Attentional focus theories were applied to everything from optimizing learning to developing self-esteem, even to performance in bed. Experts were tuning in, and the rest of us were tuning out.

Zooming In Versus Out

Have you ever been so locked in, so focused on the task at hand that you haven't noticed your spouse calling your name or maybe even the phone buzzing and beeping in the background? Often, we try to create such a state of blissful unawareness to the world around us when we're working on our craft or deep in the throes of writing. We intentionally fixate on what we're doing, neglecting everything else. Other times such a state is thrust upon us. When pressure or stress is high, our world narrows in on what's right in front of us. When airplane pilots perform a nerve-racking landing amid heavy crosswinds, they shift from the broad focus of being aware of the array of dials and instruments scattered around the cockpit to zooming in, seeing only what's right in front of them. In one study, 40 percent of pilots missed a loud beeping alarm during a wind-shear landing. When our attention narrows, our cognition follows. Our eyes stop dancing from side to side and become stuck, staring at a few small points in front of us. All that matters is the narrow band of the world. Everything else is shut out. Pilots call this inattentional deafness. Scientists call it cognitive narrowing.

There are advantages to such a response. When we narrow, we're redirecting all of our computing power to the handful of processes that matter. It's as if to help with our sluggish Wi-Fi, we disconnect our phone and tablet, just so that our video conference call won't lag. Narrowing also helps with goal attainment. It cuts out all of the other distractions and places the most important goal front and center. When we home in, we increase motivational intensity, reinforcing that what's in front of us is what we should be after. For a brief moment, the trade-off can be worthwhile, but when we remain zoomed in for too long, we start to miss cues and signals. We get locked in on one path without being able to step back and see a better route. When we're stuck narrowed in for too long, accidents go up and performance drops. We miss hearing alarms that signal there's a problem elsewhere.

If marathoners are zooming in and pilots need to do the opposite to avert disaster, what do we do if we are on the brink of an argument with our spouse or a breakdown from the demands of our job? Do we copy the pilots or the runners, or take a different tack altogether? How do you know when to zoom in and when to zoom out in a challenging situation that risks spiraling out of control? New research helps provide an answer.

How do pilots snap out of inattentional deafness, come back to normal, and pay attention to the dozens of sounds, sights, and signals bombarding their sensory centers? Do the opposite: zoom out. When researchers out of the University of Michigan strategically reminded participants to broaden their attention throughout a stressful landing, their performance improved markedly. When we broaden our cognitive scope, it's as if we take the blinders off, open up the world to include our periphery. Going broad prevents us from getting too locked in on a particular decision or action. It also changes how our mind works.

Imagine you are looking at a picture of a Rottweiler stretched out on a couch, eating a donut, and you've got one minute to come up with a title for the photo. What's the most unusual bird you can name off the top of your head? Think of as many functions of a brick as you can. These bizarre questions are examples of how scientists test for creativity. The more unusual the answers, the better. If you could only think about using a brick as a building block for a wall, your creative juices are pretty low. How about using a brick to break into your locked car? A tad better. If you answered that you'd grind up the brick into minuscule pieces and then create a brand-new style of makeup out of it, your decision making might be questioned, but your creativity wouldn't.

Creativity is key to zooming out because it is the cornerstone of problem-solving, which is essential to our rethinking of toughness. Creativity broadens our world, opening up potential paths and preventing us from defaulting to the well-worn route that may be easy to follow, but ultimately leads us to more frustration. Whether you are facing your twentieth rejection letter for that book that deserves publication or trying to keep your sanity while wrangling twenty five-year-olds in a classroom, a little imagination can be the difference between giving up and finding a way forward.

When researchers asked similar questions to a group of University of Maryland students, they weren't concerned with classifying who was creative and who was dull. They wanted to see if they could prime people's creativity. To get participants' imagination flowing, they handed them a map of Arkansas—not exactly a beacon of inspiration. They instructed half of the participants to look over the entire state. For the other half, they highlighted a bright-red star in the center of the map, representing the city of Little Rock. This simple task forced people to zoom: narrow their focus on the city, or broaden their focus to see the entire state.

Focus on the tree or see the forest. Afterward, they were handed a piece of paper with one of the aforementioned creativity puzzles to solve.

The students primed with a broad attention task produced more unique answers. Those who focused on the city of Little Rock weren't as fortunate. Their minds were closer to bricks as building blocks instead of makeup. Variations of this experiment have been conducted using a variety of similar tactics, such as focusing on a leaf of a plant (i.e., narrow focus) versus looking at the plant as a whole. Regardless of the setup, the results consistently show that when we go broad with our attention, our imagination runs wild and our creative juices flow.

A narrow focus doesn't just leave our imagination empty; it plays a large role in rumination and depression. Think back to chapter 7, where we discussed how our inner dialogue could spiral out of control toward a freak-out moment. Our thoughts shift toward the negative, and all of the feedback we receive seems to validate our experience. Before we know it, we've quit whatever task we were performing, whether it is working on the same paper for nine hours without taking a break or trying to assemble your kid's playhouse for three hours using the bizarre disposable tools that came with it. That narrow focus blinded you to even considering other options or tools—and it made you either furious or drained, or both. One group of psychologists believe that rumination occurs when we have a narrow and inflexible focus of attention. We become so focused on a singular aspect (e.g., a comment our boss or coworker made) that we can't break away from that thought or sensation. In the case of rumination, our narrow focus primes us to amplify the negative thoughts and doubts circling in our mind. Soon enough, we're incapable of seeing or listening to anything but that singular voice. Suzy's snide remark about our presentation is stuck on replay in our mind. We can't pick up or listen to any sensory information that may let us know that our doubts are unfounded. Our worldview shrinks to what's right in front of us, screaming the loudest. Or put another way, if we are asked what a brick could be used for, our minds can't reach beyond its function in building a wall.

Psychologist Barbara Fredrickson proposed a similar framework that works for emotions. The broaden-and-build theory of emotions states that positive emotions expand our cognition and our opportunities for action. According to Fredrickson, when we experience positive emotions, we're more likely to have novel thoughts, take on new challenges, and embrace new experiences. On the other hand, negative emotions tend to make us narrow our possibilities. Negative emotions constrain our thoughts and behavior. Our options become limited when we're swamped by anger. Whether it's attention, cognition, or emotion, the pattern is clear. Broad is the way to go; narrow is to be avoided.

How do we make sense of this broad-versus-narrow information in the context of the research on Frank Shorter and his world-class pals who tended to use associative strategies, which are narrow by nature? Focusing on how our arm is swinging, or the depth of our breathing, or the sensation of fatigue in our thighs? And isn't dissociation, the preferred choice of slower runners, similar to a broad focus? If zooming in is tied to ruminative thoughts and a declining mood, then how do we make sense of the fact that some of the top marathoners in history zoomed in? If it seems like contrasting information, it's because it is.

Adjusting Your State of Mind

Every second we are bombarded with information from within our body and from the outside world. Do we wait to react to it or act before we're fully aware of what's going on? When it comes to making sense of the world and the information assailing us every second, our brain relies on two types of processing: top-down and bottom-up. The former relies on the brain acting in a predictive manner, utilizing experience and expectations to predict what will occur. Top-down processing is context-driven. It's present when you feel a wave of panic and anxiety take over when you are sitting quietly, waiting to be called up to the podium to give your speech. You haven't seen the audience or even the stage, yet based on context and prior experience, your brain is preparing for what's likely coming. On the other hand, bottom-up processing is sensory-information-driven. You pick up a cue that triggers a reaction. Your brain is working in real time, reading the information and then mounting a response.

These processes aren't mutually exclusive. It's not as if our brain uses all top-down or all bottom-up processing. In most situations, we're using a combination of both. A dash of prediction, a touch of feedback, making our best guess using context, and taking in sensory information to either direct or course-correct our prediction. In 2020, neuroscientists Noa Herz, Moshe Bar, and Shira Baror proposed that at any given time, we fall somewhere on a continuum between the two extremes of all top-down and all bottom-up processing. Where we fall on that continuum impacts not only our perception, but also our attention, thoughts, mood, and behavior. All are connected to either more top-down or more bottom-up processing. Herz and colleagues chose names for the top-down and bottom-up sides of the spectrum: narrow and broad.

The type of processing is the glue that binds perception, attention, thought, action, and mood together. Move toward the broad side, and we shift toward relying on sensory information for perception, having global attention, broad thinking, exploratory behavior, and a positive mood. Move toward a narrow state of mind, and our focus shifts to the trees instead of the forest: our thinking becomes constrained, we tend to fall back on familiar choices and actions, and our mood sours. We linger where we are instead of exploring.

The beauty of the state of minds theory is that it provides a simple framework founded on deep science to explain why zooming works. It explains why priming our mind to go broad leads to more creativity. It's not just shifting our attention so that our vision picks up more information. It's that we're shifting how our brain functions. Dragging it from the narrow top-down function to the broad bottom-up. And as we change the ratio of processing, our attention, thoughts, action, and mood get dragged along. Narrow our attention, and our thoughts and mood follow suit. Broaden our thinking by performing a task that requires thinking holistically, and the rest gets dragged with it. In other words, seeing the forest instead of just the trees might not just help as a metaphor for life, but it could also change your current mood.

The theory also explains why the old model of toughness fails: it limits our path. In the old model, to move forward, we have to dig in, grit it out, play through the pain. And if that fails, we're limited to doubling down. Going through the same experience, hoping that somehow that outcome is different. The new model works much in the same way as the state of minds theory. It's about shifting how our brain works, and in turn how we interact with the thoughts, emotions, and feelings that come with a challenge. Instead of doubling down, we open ourselves up to new paths. Sometimes going broad, other times narrow.

Experiencing rumination? You've got narrow thinking, narrow attention, and constrained behavioral response. Attention, thought, and action all line up to lock into a cycle that is difficult to get out of. Throughout this book, we've detailed experiences where we spiral: we ruminate on an issue, our world narrows, our mood shifts, and if Herz and colleagues are right, we've shifted to where we are almost all the way to the extreme of top-down processing. Stuck in a loop where the only thing we can predict is doom.

How do we get out of this narrow spiral? No different than the pilots who needed to broaden their view to get out of inattentional deafness. Or the science that tells us to zoom out and think in third person instead of first person to broaden our perspective and disconnect ourself from the experience. If we find ourselves getting narrow, counteract by going broad.

This theory helps explain the advice that "Mood follows action." When we're feeling sad and down, trying to alter our mood seldom works. But if instead of trying to force your mood to change you change your behavior—getting out of bed and going for a walk—you often find yourself in a much better, happier place. Your behavior dragged your mood with it. Rich Roll, a well-known podcaster who struggled with drug and alcohol addiction in his twenties and thirties, explained to me how he utilizes this principle: "If I'm down or in a rut, I force myself to move my body, even if only a little bit. This helps shift my perspective and reset my operating system—and more often than not, the sun starts shining again."

It's not that having a broad or narrow state of mind is better or worse. A narrow mind can be beneficial, keeping us focused on what's right in front of us, throwing all of our resources behind predicting what's going to happen next, and preparing for how to handle it. Zooming in allows us to see the finer-grained details, to keep our goal front of mind, and to resist the pull of outside forces trying to steer us away from our current trajectory. The moments when we need to be on the far end of narrow processing are generally very short. If we linger, we spiral. Zooming out allows us to take in more information, to see new patterns and connections between disparate ideas. We're able to explore, to quickly shift the direction we're headed. But, if we get stuck in a broad state of mind, we are left reacting to the environment as it comes. We spend too much time exploring the world around us and not enough taking advantage of the information and action possibilities sitting before us. Top performers figure out when to go broad and when to narrow. And that's the secret Frank Shorter and his marathon pals had stumbled upon.

Exercises: Broad Versus Narrow: Changing Your Processing Ratio
The first step in utilizing any of the strategies below is to decide whether you need to zoom in or zoom out. Remember, stress causes us to narrow because it's advantageous for a short period of time. Stress locks us in and shifts us to focus on one goal instead of exploring others. But you also miss vital information, and over time, your inner voice and negative mood soon spiral. Elite performers can zoom in for longer than novices. They get the benefits of locking in on a goal without the negative mood and rumination following too quickly. It's about being able to flexibly zoom in and out as needed. Remember that each strategy is designed to drag the other categories with them. If we shift our attention, we're pulling our thoughts, actions, and mood with us.

1. Visual Zooming: Portrait Versus Panorama Mode
Direct your focus of attention, almost staring at an object, picking up as many details as possible in a small area. That's portrait mode. It primes your mind for a single task, shifting you to a narrow state of mind. On the other hand, softening your gaze to an almost blurry state where you attempt to pick up everything in the periphery is what I call panorama mode. You want to broaden your attention. When feeling overwhelmed by discomfort, taking

a second to blur your vision helps shift you away from spiraling downward.

2. Cognitive Zooming: Weird Versus Normal
This is what I like to call the *Family Feud* style of thinking. In the TV game show, they present you with a challenge (e.g., "Name something that goes up and down"), and then you have to answer what you think people who were surveyed would have answered. Early on, you want to guess the most common answers, like maybe an elevator. But once you're down to your last answer on the board, you have to start guessing strange things that maybe one or two random people on the street might have come up with. That's narrow versus broad thinking. Common answers are narrow; unusual answers are broad. The former primes us for focus and concentration, the latter for creativity and innovation.

3. Physical Zooming: Mood Follows Action
We have covered this idea so far in the book, but to reiterate, one study took participants and sat them in a chair. They told participants to either lean forward so that they were on the edge of their seat, anticipating what was coming, or lean back in a fully reclined comfortable position. After getting into position, subjects were given a task to categorize a group of pictures. Those who reclined in the chair were more likely to choose broad categories, coming up with creative ways to make, say, a vehicle and a camel fit in the same category. When on the edge of their seat, they stuck to narrow categories. Not only does our mood follow action, but so do our thinking, perception, and so forth. Alter your actions to go broad or narrow.

4. Temporal Zooming: Imagine the Future
When you are going through a difficult time, ask how you might

feel about this in six months, a year, or even ten years down the line. Imagining how future you would think about your current situation often reminds us that whatever we are going through is temporary. And that in the future, it'll be an indistinguishable blip in our life story.

5. Linguistic Zooming

Remember from chapter 7 that when we switch from first person to either second or third person, we are putting distance between what's occurring and our response to it. This doesn't just apply to our inner dialogue, but journaling using second or third person can also help us process our emotions.

6. Environmental Zooming

To get much work done, modern writers often have to find a quiet desk tucked away in an office with little to no distractions and, of course, the Wi-Fi disabled. They want to narrow in on the task at hand, so they create an environment that allows them to do so. On the other hand, research shows that going for a walk out in nature, or even looking at pictures of awe-inspiring natural wonders, sparks creativity, shifts us so that we see difficult situations as a challenge instead of a threat, and allows us to bounce back and recover more quickly from stress. Nature expands our perspective. Set up your environment so it invites the action and mental state that allows you to perform.

To Suppress or Address: How to Navigate Life and Death

In the fall of 2010, Katie Arnold welcomed her second daughter into the world. Three months later, she watched her father succumb to cancer. Reflecting on this challenging time, Arnold told me, "It was a perfect storm of grief and postpartum. . . . I was in a kind of hormonal stew with everything turning topsy-turvy." For the *Outside* magazine editor, the whirlwind of life's giving and taking set off a cascade of emotions. "Love, fear, rage, regret, disappointment, tenderness, shame, surprise, anguish, even awe. Grief is all of these things and more; a big, messy wad of emotion. It is beyond category, as fathomless as desire, as luminous as joy. It will break your heart and fill it up again," Arnold wrote in her memoir, *Running Home*. The anxiety was immobilizing and crippling. To Arnold, grief wasn't just an emotion; there was a physicality to it. She explained, "I knew that grief was an emotional state, but I didn't know it was physical. My grief manifested as pain in my body—aches in my joints, this heavy weight on me—and I became afraid that was a signal that physically something was wrong with me. . . . I had a conviction or certainty that I was dying too." Any ache jumped straight from slight discomfort to a full-blown fear that the pain in her elbow or stomach or knee must be cancer. This went on for a year and a half.

There are few experiences as universal as dealing with the mess of emotions that comes with death and figuring a way through it. Unbeknownst to Arnold at the time, her experience of grief as something physical and her taking on the pain of the loved one she lost are relatively common when going through grief. Arnold was at a loss. "I could not talk my way out of it. People could not reassure me this was not true. . . . I tried lots of different healing and therapies, some worked a little, others did not, but really what worked for me was running. Moving on my own two feet through the wilderness," she relayed. Being alone in her head "was the only time I could move beyond my fear thoughts. At the start of every run, my anxiety was super heightened, but as I ran, that rhythmic repetition lulled me into a moving meditation where the thoughts dropped away. And a realization that the body was strong. The body holds the wisdom, it was sending a message, 'Actually, you're healthy, Katie.'"

Running became her salvation, her time to quiet her mind and the fear and rumination that came with it. There wasn't competitiveness to it, as she told me: "It was a meditative, thought-free place where my imagination wasn't going haywire." The mind-clearing exercise eventually led the once-recreational runner to the top of the ultra-endurance world, where she topped the women's field of the prestigious Leadville 100-miler in 2018.

One of the hardest things that anyone can go through is losing a parent, spouse, or child. The grief can be overwhelming, pushing us toward depression and despair. When the death of a loved one occurs, we're left with a gaping hole in our lives and a mess of emotions that we can't quite make sense of. Some of us bottle up our feelings or distract ourselves by going all in on our job to numb the pain. Others choose to work through the grief, talking to a friend or therapist about what they are feeling. Managing grief often feels like our own internal ultra-marathon. Beyond the question of zooming in or out as a coping strategy, how do we deal with and regulate our emotions, whether those are grief, sorrow, or even fear? Do we ignore them, move on, embrace them, or fall prey to them?

There are several different ways to regulate our emotions. Broadly we can categorize them based on whether they rely primarily on attention or cognition. Attention strategies direct our focus toward or away from the emotion. We can concentrate on an item, shining a spotlight that amplifies the sensations around it. Or we can distract ourselves, filling our working memory with something else to deal with, preventing that initial feeling of anger from having time to grow. We focus on our children or our work, to keep our mind from drifting back to our loss. Distraction and concentration are low-cost strategies, requiring just a smidge of cognitive effort—a simple shifting of where we shine our spotlight. We're not trying to do anything with the actual experience.

Cognitive strategies are more complex, utilizing cognitive control to suppress, detach, or reappraise the emotion. Instead of relying only on directing our focus, cognitive strategies involve actively engaging with the emotion. Suppression relies on dampening down the feeling or sensation, while reappraisal involves reframing the experience into something that you can handle.

If psychologists could crack the code on what strategies allowed us to deal with sadness, for example, the benefits to humanity would be enormous. Like prescribing a rehab exercise to cure that old knee pain, we could prescribe different coping strategies to deal with the emotional turmoil in our lives. As research mounted in the late twentieth century, scientists went as far as labeling strategies like concentration that led to rumination as maladaptive, while others like reappraisal were seen as adaptive. And the research backed it up. When subjects concentrated on their thoughts for too long, their emotions spiraled. As a strategy to handle sadness or grief, rumination failed repeatedly. Reappraisal had the opposite effect, helping shift individuals' experience from negative to positive. Turning anxiety into excitement, or disappointment into an opportunity for growth, seemed to help people deal with failure or grief in a productive manner. Some strategies helped and others hindered.

But like most aspects in life, as we dive deeper, the straightforward narrative disappears. Starting in 2010, it became apparent that the good-and-bad dichotomy fell apart on an individual level. It wasn't as simple as feeling anger, then trying to distract ourselves. Emotional regulation required more nuance. In a 2011 study titled "Emotion-Regulation Choice," a team of scientists led by Gal Sheppes and James Gross at Stanford signaled a need for a course correction. They stated, "These studies have enormously advanced the field of emotion regulation. However, a new generation of studies has begun to cast doubt on the unconditional maladaptive/adaptive label given to different strategies." After the alarm was set off, a series of studies validated what Gross and colleagues suggested: good and bad was too simple. Everything worked—and failed. Every strategy had benefits—and came with a cost.

Take, for example, the principal maladaptive strategy of rumination. It's easy to see why letting your thoughts spiral hinders emotional regulation. After all, rumination has a close relationship with anxiety and depression. But researchers have found that rumination helps when we have to maintain focus on a singular goal. If our mind and thoughts get stuck on a goal that matters and is beneficial, rumination keeps us from drifting away. So-called deliberate rumination causes us to make fewer errors when performing singularly focused tasks, and actually facilitates post-traumatic growth after a harrowing experience. In tasks that require shifting between different goals, it largely fails.

Another strategy that *typically* makes matters worse is suppression. What inevitably happens is when we actively resist, the object of our ire comes back with a vengeance. However, researchers found that suppression works well when the situation is particularly dire. For instance, suppressing negative emotions after a spouse passes away helps reduce grief over the long haul. Compartmentalizing is a form of suppression that many of us use in our everyday lives. We note a sensation of stress but tuck it away in the corner of our mind until we have the resources to address it. That latter part is the key. Suppression is a short-term solution. Eventually we have to deal with the experience.

Even reappraisal, a strategy that a meta-analysis of nearly two hundred studies showed was the most effective coping strategy, fails in some instances. When dealing with intense emotional situations, our capacity to focus on and then reinterpret what an emotion means falters. When compared to strategies like distraction, reappraisal requires more cognitive bandwidth. Instead of directing your attention elsewhere, it requires focusing on an item without being overpowered by the feeling or thought, and then devoting a large chunk of your mind's resources to altering your initial view, all while under a heavy load of stress or fatigue. Reappraising anxiety as excitement is easy when the feeling comes from getting up the nerve to sell Girl Scout Cookies to strangers; it's much more difficult when that sales pitch is to a boardroom full of investors and your company's future is on the line. Even the best strategies fail when we don't have the capacity to use or handle them.

If everything works—and doesn't—how do we make sense of and apply all of this information when we are faced with something difficult? Research on bereavement provides an answer. In one study, psychologists set out to understand why some people could cope with losing a loved one reasonably well and why it sent others careening into a long period of despair. To understand what was going on, they took forty individuals who had lost a spouse in the past three years and had them evaluated by a number of clinical psychologists. The subjects were then classified as either handling their loss with some success or going through complicated grief: an experience of prolonged loneliness, difficulty in accepting what had occurred, or a sense of the meaninglessness of life. Once classified, both groups were put through what's called an expressive flexibility task. Subjects are presented with a series of pictures designed to evoke a range of positive and negative emotions. As images of cute puppies, the World Trade Center tumbling down, or feces smeared all over a toilet popped onto their screens, participants were asked to enhance, suppress, or do nothing with the emotion the picture evoked. The better they were able to turn up or down the volume on the emotion, the better their expressive flexibility.

Not surprisingly, those who suffered from complicated grief scored lower in expressive flexibility. When it came to regulating or adjusting their emotions, they were stuck. Incapable of adjusting the inner volume. The successful grievers, on the other hand, had control over the experience. They could amplify or suppress their emotions, no matter what was presented. Coping flexibility allows for adapting to adversity in a positive direction over the long haul.

The best performers tend to have a flexible and adaptive coping ability. They can bounce between different strategies, depending on the demands of the situation. Top marathoners possess this ability, as do the most resilient individuals who overcome grief and trauma. Flexible coping is tied to everything from dealing with trauma to college students adjusting to life on their own and homesickness. Sheppes and Gross offered the following summary: "Healthy adaptation is the result of flexibly choosing between regulation strategies to adapt to differing situational demands." It's not that distraction, suppression, reappraisal, detachment, or shifting to a broad or narrow worldview is good or bad. They all work—and don't.

TOUGHNESS MAXIM

The best performers tend to have a flexible and adaptive coping ability. They can bounce between different strategies, depending on the demands of the situation.

Flexible and Adaptable

In 2017, I conducted an informal study comparing where the minds of a group of college and professional athletes went during discomfort. During a series of grueling workouts and competitions, the athletes completed a survey about where they directed their attention. Contrary to the earlier research, the best athletes weren't relying on association and poorer performers weren't relying on dissociation. Everyone was using both strategies, just to different degrees and at different times.

The best athletes reported tuning in to their body at crucial moments, getting the lay of the land on how their body felt. At other times, they reported trying to "zone out." They even had a name for the strategy after it was utilized with great success: turning your brain off. One athlete described the experience like this: "I tried to turn my brain off, lock onto the competitor ahead of me, get in autopilot for a bit, just letting my body do what it knows how to do. Until the race got more difficult, and I'd need to zone back in. It was like I was storing up my energy until I really needed to dig deep."

They weren't just using one strategy; they were changing their focus at different times in response to the demands of the event. Fatigue and a rising uncertainty about whether they could finish? Flip the switch and home in. Are those around them starting to make tactical moves? Shift attention to their surroundings and competitors. As I quizzed the athletes, the best performers were directing attention to help cope with the demands of the race, and the strategies were vast and complex. It wasn't as simple as associate or dissociate. And recent research backs up my observations.

Since Pollock and Morgan's seminal study in 1975, a clearer picture has emerged. At the 1988 Olympic Trials, John Silva and Mark Appelbaum interviewed thirty-two of the United States' top marathoners and found that while association was the dominant strategy used, dissociation was used as well. The top marathoners reported being "in flux," changing between strategies. Silva and Appelbaum concluded that elite marathoners employed what they called an adaptive flexible strategy. Now we can fully understand Frank Shorter and his pals' approach to mastering the discomfort of the marathon.

Elite runners obtain this flexibility by developing their capacity to utilize strategies while under stress and fatigue. The reason novice marathoners default to dissociation is that they quickly become overwhelmed when the intensity of the experience gets too high. They don't have the capacity to use strategies that take a large amount of cognitive effort or that might amplify the experience in the short term. Their resources are directed at surviving. When emotional intensity gets too high, we default toward the easiest path: distraction. It's a low-cost solution that allows us to manage something in the short term.

On the other hand, during periods of pain, fatigue, and suffering (i.e., high emotional intensity), elite runners can still use strategies that require directing attention toward the experience. They can focus inward on a feeling, thought, or sensation without it overwhelming them. The ability to home in, when their mind is screaming at them to turn their attention elsewhere, allows them to extract more information, and potentially reappraise what they are experiencing. Without the ability to focus, reappraisal can't occur. Elite runners don't just choose to associate more than novices. They have the capabilities to do so. Through years of training, they've learned to use regulation strategies that take more resources and effort, even during the most demanding moments.

This particular skill isn't reserved for the fastest runners among us. It's a pattern that we all experience. As any parent can attest, toddlers and young children are prone to emotional outbursts filled with crying, tantrums, and fits. They are masters at generating strong emotions, but novices at regulating. Their skill set is heavily tilted toward emotional generation, and it shows. Part of development and getting through the terrible twos and into the school-age years is tilting that seesaw back in favor of emotion regulation.

In tracking how children develop the capacity to regulate emotions, early-stage strategies surface first. Infants as young as six months old start to show signs of attention regulation. They divert their eyes from objects that cause them distress. By the time they turn two, distraction seems to be the coping strategy of choice. Two-year-olds are capable of turning their attention to another object or person to deal with emotional discomfort. Reappraisal and other cognitive strategies aren't developed until much later. In two different studies, researchers measured the brain's electrical activity to see whether children could successfully reappraise an emotional stimulus. When we are able to reappraise a negative emotion and turn it into a positive one, there's a reduction in the late positive potential (LPP), a neural signal that reflects how much attention we're giving an emotion. When researchers studied five- to seven-year-old children, there was no change in electrical signal, indicating that children hadn't been able to reappraise fear and anxiety. However, in a study on eight- to twelve-year-olds, researchers found a change in LPP, indicating a successful reappraisal. Whether children learn to reappraise at six or eight isn't the point. It's that cognitive strategies like reappraisal take time to develop. They are skills. In many ways, our novice marathoners are like the young children capable of distraction, while our elite runners have the full capacity to use more cognitively demanding and complex strategies.

At one time, we were all the child who moved from annoyance to despair, as we struggled to regulate our inner turmoil. Our journey toward real toughness is much the same. We start with only one solution: grind it out, ignore much of what's going on inside. It may help us get through minor difficulties, but eventually it fails. We have to adopt a way to navigate the complex feelings, emotions, and thoughts that swirl whenever we are faced with a challenge. Over time, we gain the ability to pay attention to our inner world, navigate the experience without freaking out, and ultimately make better decisions. As adults, we all have the machinery and capacity to develop that ability.

The Capacity to Cope

Imagine for a second you're standing with a few friends at the top of a snow-covered hill. One of your friends decides it would be a good idea to make a nice round snowball and send it careening down the hill. After a gentle nudge, the snowball starts to pick up steam and size. As it makes its way down the slope, the bigger and bigger it grows, and the faster and faster it goes. You glance down to the bottom of the hill, and a family is sitting there quietly, unaware of the now-giant snowball barreling toward them. What have you done? By the time you notice the family, the snowball has gained so much momentum that it's nearly impossible to stop it. You can't get in its way, and you can't divert it. Welcome to a full-blown freak-out.

We can think of the process of toughness as the snowball on the hill. Our emotions, thoughts, and feelings work in the same way, gaining momentum as we make our way from initial sensation to taking action. If you've ever found yourself ruminating on an argument, replaying it over and over, you are familiar with the momentum-gaining power of feelings, emotions, and thoughts. According to the process model of emotions, emotions gain steam as they go through cycles of attention, appraisal, and response. We start with a small, simple feeling, but as it stirs around in our mind, as we divert more and more of our attentional resources, as we try to react to the discomfort, it grows in power. We've pushed the snowball down a hill. This cycle repeats until the emotion has been extinguished, been redirected, or taken over.

Whenever we face a situation that requires toughness, we are attempting to stop the snowball careening down the mountain. If our only hope to stop the snowball is to stand in its way, put our hands up, and yell, "Come at me!" chances are slim that everything will turn out all right. That's old toughness. A simplistic solution to a difficult situation. Real toughness is about having a myriad of potential options. Many will fail, but the chances are a lot higher that we'll find a way to stop or minimize the damage of a careening snowball.

We can intervene early, putting a stop to the ball before it even starts to roll. Or we may be able to slow the momentum of it, making sure that it's only nudged instead of shoved. That's akin to using distraction, a low-cost strategy that works well early before it's gained too much momentum. But what if you can't catch the snowball before it has picked up speed? That's where cognitive strategies like reappraisal come in. Instead of trying to stop the snowball in its tracks, you are trying to push it onto a new path.

Being tough is standing on that hill, recognizing the danger, and figuring out how to keep that snowball from gaining speed and crashing down into the family standing at the bottom of the hill. If we can take away its power by slowing its momentum, and if we have the ability and flexibility to use one of several options at any point along the journey, then our chances of successfully intervening are higher.

When it comes to developing toughness, the role of coping

strategies is clear. Our ability to be tough is dependent on the emotions and thoughts that push or pull us toward a decision or behavior. Coping strategies act to amplify or blunt the effects of those thoughts, feelings, and emotions. We can use simple techniques, like directing our attention, or complex ones that involve cognition, like reappraisal. Where we direct our focus and cognition can push us toward a freak-out or enable us to work through whatever it is we are facing.

The old model of toughness—ignore and suppress—pushed us toward a limited selection of strategies. We were taught that experiencing doubts and fears was a sign of weakness, that we shouldn't listen to the sensations of pain or fatigue that were screaming at us. In the old model, the preferred way through this jumbled mess of thoughts, emotions, and feelings was to bulldoze our way to the finish line. That approach ultimately backfires.

Thankfully, modern research and practice teach us another lesson: develop an arsenal of strategies so that when adversity arises, you'll have the right tool for the job. Be adaptable instead of rigid and stoic. When it comes to using coping strategies, the science points to two key attributes that we need to develop:

- Flexibility to use different strategies
- Capacity to be able to utilize them

When it comes to making tough decisions, we need to practice different strategies: zooming in and out, distracting and associating, and so forth. The labels and categories don't matter as much as the ability to use them. It's only through applying the different strategies in a variety of situations that we learn what works. And sometimes, your tried-and-true way of dealing with a thought or emotion simply won't work. As world-class runner Brian Barraza told me, "You cycle through different strategies, seeing what works and discarding what doesn't. Over time, you pick up that certain strategies work better in certain races or situations. But if you build your arsenal, what you've got is two to three different strategies for each point in a race." Practice and experience teach you what works when.

The Other State of Top Performance

"That didn't even feel hard!" the athlete exclaimed after achieving a personal best. Diving further into their experience, they rattled off how seemingly magical the performance was. "I was in the zone. Everything was clicking. It was like I was just watching myself perform.... I wasn't even thinking about the race.... I just got into a rhythm and let my body go to work.... I was on autopilot."

When we perform on the field, on stage, or in the boardroom, we often expect it to be difficult. A race should hurt. A speech should be nerve-racking. But every once in a while, we have a surreal experience where it all comes together, and the difficult becomes routine. Renowned psychologist Mihaly Csikszentmihalyi gave this rare but welcomed experience a name: flow. According to Csikszentmihalyi, flow is a state where "the ego falls away. Time flies. Every action, movement, and thought follows inevitably from the previous one, like playing jazz. Your whole being is involved, and you're using your skills to the utmost." While the somewhat mysterious concept of flow is difficult to describe, for anyone who has experienced it, they know exactly what it feels like.

Once experienced, the state of flow becomes like a drug, something that performers seek out. Papers and books have been written about rigging your environment and mindset to increase your chances of entering this vaunted state. Author Steven Kotler founded the Flow Genome Project to discover how to unlock the secrets of flow and help us all perform better by entering the state more often. For the past few decades, sports psychology has focused on how to help performers enter a flow state. That was the key to performance. Get there more often, and great performances would follow.

After an incredible performance, individuals occasionally report the exact opposite experience to the ease and magic of flow: "That was the hardest thing I've done in my life. It was a grind the whole time. Somehow I got it done." A competition where every step required immense effort, where doubt, pain, and fatigue were at their highest, and nothing seemed to come easy. Yet, they found a way through the difficulty and came away rewarded with a win or personal best. When it comes to performing at our peak capacity, we often think there's only one path. Traditionally, that route has been through flow. But what the experience of the best performers—and the latest psychology—shows is that there's more than one road to the top.

When Christian Swann and colleagues interviewed athletes from different sports on their experience after a big win or personal best, they identified two distinct states that led to top performances. The first they described as "letting it happen," an experience that corresponded to the concept of flow. It was easy and enjoyable. The other state was much more difficult, described as gritty and grinding. This performance state occurred during an important part of the game, when pressure was high, and an outcome was on the line. The athletes were able to raise their performance not by going into autopilot but by narrowing their focus, deliberately increasing their effort, and making it happen. The athletes were experiencing a clutch state.

Two different psychological states. One easy, one very difficult. One that came to them, one they seemingly had to force. Both bringing about peak performance. When scientists set out to unravel these two states that led to equally impressive performances, a few differences emerged. While both states shared a feeling of enhanced motivation, perceived control, absorption, and confidence, they diverged when it came to three key components: attention, arousal, and effort. Flow states contained effortless attention, optimal arousal, and an automatic/effortless experience, while clutch states contained complete and deliberate focus, heightened awareness and arousal, and an intense effort.

While Csikszentmihalyi emphasized the role of complete focus in his original definition of flow, it's not the deliberate focus that we think about. One basketball player described the experience as "just really focused without being focused. . . . You're tuned in without having to focus, without having to tell yourself to focus." In one review, researchers went so far as stating, "Prolonged effortless concentration of attention is the principal characteristic of the flow experience." Attention isn't just a descriptor, but it's also a way to facilitate or maintain the experience. In another study, Swann and colleagues found that top athletes reported using distractions, such as daydreaming or letting their mind wander off to the scenery around them, as a tactic to stay in a flow state. Similarly, golfers who reported being in the zone deliberately directed their focus away from the game, talking to their caddie so that they didn't get in their own way.

On the other hand, to experience a clutch state, athletes report narrowing their focus instead of zooming out or utilizing distraction. They shift their goals from being open-ended (e.g., "I want to win") to being fixed and specific ("I need to score five points in the next minute"). They heighten their focus to amplify their arousal, homing in on the feeling of nervousness to keep them from being complacent. Or attending to the pain so that they can make a quick assessment of what they have left in the tank and how to use it. Narrow, zoomed-in, focused. A deliberate use of attention to get them through tough times. But that wasn't all. Athletes experiencing clutch states report one more difference. They had to actively make a decision.

"There was a definite feeling of a switching of gears, and 'Right, okay, things are quite serious here.' . . . A feeling of having to take action," a polar explorer reported to Swann and colleagues when they were investigating clutch states. Clutch states didn't just appear. Athletes didn't just happen upon them. With clutch states, there was a conscious decision to increase effort and intensity. They had to flip the switch. How they did so varied, but when they were at the toughest part of the performance, they figured out how to *choose* to increase their effort. Clutch required choosing; flow required experiencing.

Two different states. Both bringing about top performance. One requires grit, the other grace. One accepting, the other a conscious decision. In many ways, the clutch-versus-flow paradigm reflects toughness. We tend to think of it as a singular method: push through, persist. But as we've come to realize, that's a false constriction. Being tough means being able to choose the right strategy, given your abilities and the situation. How do we square the push for equanimity and the flexibility to suppress, ignore, accept, or redirect as needed?

In discussing growing from adversity in his book *Transcend: The New Science of Self-Actualization*, psychologist Scott Barry Kaufman stated that the key is to "have equanimity as your default, but to retain the capacity for defense, fighting, and taking a forceful stand." Equanimity is about creating the space to be able to respond. And sometimes, that means choosing to flip the switch.

THE FOURTH PILLAR OF TOUGHNESS

TRANSCEND DISCOMFORT

CHAPTER 10

Build the Foundation to Do Hard Things

"Don't talk to them. They're your competitor, your enemies," the coach sternly commanded during the drive to the stadium. This was confusing to hear for Julie, a college freshman and newcomer in her first competition.* After all, a few months earlier in high school, she'd considered many on the opposing team friends. She'd competed against some of them back in junior high school and had shared meals, pregame chitchat, and even long bus rides. A few had even become her training partners, and they'd formed a small workout group during the off-season when high school practices weren't allowed. Now, all of a sudden, they were enemies? The veterans on the team seemed to follow through, shunning anyone but their teammates on race day. Julie felt terrible but went along with it.

As the season progressed, Julie noticed a few peculiar behaviors from her coach. She would pour praise on a few athletes while seemingly ignoring others. At one point, Julie found herself in the doghouse, desperate for some positive feedback and willing to do just about whatever the coach told her at practice to receive some. The coach even seemed to pit her team members against one another. First, there was the list posted on the coach's office door,

---

* This experience is an amalgamation of athletes' experience at the collegiate and professional level, with names altered to protect their privacy.

which ranked everyone on the team from top to bottom, not based on any objective metric, but more on who was the coach's favorite of the moment.

Second, there were the stinging comments and invasion of privacy. The snide remarks about Julie's weight and the questions about her personal life. "Sex makes you gain weight," the coach remarked at one point, a strange comment for a twenty-something to hear from a fifty-something. Looking back, Julie almost laughs while describing it all. "It's so strange and absurd that it's almost comical."

But at the time, it felt normal. Here was this authority figure that had won championships. Everyone fell in line. They all bought that this was the path to reach their goals—that everyone on the outside was out to get them, that the only people they could trust were in their inner circle. Or, as her coach once told them, "It's us against the world." As Julie reflects on her college athletic experience, she pauses for a moment before uttering, "It's almost like we were in a cult."

In a strange way, a cult and a team have some startling similarities. In both, you are asking for a near-complete buy-in to a mission. You sacrifice your individual desires or needs for those of the group. And oftentimes, you develop in-group-versus-outgroup dynamics, believing that your team is better than all of your competitors. That the way in which you practice, train, and prepare is superior to all others.

One crucial difference lies in how the buy-in is established. In a cultlike environment, fear and control predominate. And it can be downright abusive. According to research, there are four key elements that controlling leaders utilize to create individuals who are dependent on them:

- Controlling use of rewards
- Negative conditional regard
- Intimidation and isolation
- Excessive personal control

When it comes to high-level performance, we're used to excusing away behaviors that would be inappropriate in everyday life as being necessary for "toughening up" individuals. As I've outlined, in sports, this view often prevails, but it also shows up in the business world. We lift up the genius of Steve Jobs and brush aside his harsh leadership style that included terrible outbursts, rash firings, and screwing over business partners, friends, and even family. As the opening sentence of a *Forbes* article succinctly put it, "Steve Jobs was a major, world-class jerk." But it worked, right? Apple transformed into a world-changing company thanks to Jobs's leadership.

We carry around the idea that sometimes people need to be pushed to their limits. That sometimes the only way to get through to them is to scream, punish, or demean. That in doing so, the individual might suffer in the short term, but in the long term they'll be thanking us. Sometimes a little discipline and demand are what someone needs to reach their potential. Occasionally, we have to force people to do really difficult things, because after all, we are inherently lazy, right?

That's the idea many of us hold on to: that something or someone from the outside needs to keep our lazy self in line. Look no further than the reason given for cutting unemployment insurance during the coronavirus pandemic. "It wouldn't be fair to use taxpayer dollars to pay more people to sit at home," stated US Secretary of Treasury Steve Mnuchin. Implying that without incentive to work, most of us would sit around doing nothing. It's a common notion, one deeply ingrained in our Protestant work ethic society. But it's wrong.

In a meta-analysis covering over 120 years of study, there was less than 2 percent overlap between salary and job satisfaction. And a Gallup study of over 1.4 million employees found no relationship between employee engagement and level of pay. A recent analysis found that people who are motivated by their inner drive are three times more engaged than those motivated by the external. We have this idea that people need to be pushed or have a carrot dangled in front of them, or else they'd all sit at home watching Netflix. That's wrong. The inner drive matters more than the outer.

The ability to persist, to stay motivated and engaged, is directly tied to toughness. How do we keep going, despite ever-increasing stress or fatigue? For a long time, the answer was centered on control and being motivated by something external. In chapter 5, we discussed how persistence is dependent on perceived control, but that only tells part of the story. Our *why* tells the rest.

When researchers looked at the ability to persist in various domains—from cycling to math to calling for donations—one factor kept popping up. Those who persisted had different goals. They weren't driven by fear, or guilt, or pressure. They didn't keep working in pursuit of money. They were pursuing a goal because it aligned with who they were and brought them enjoyment and contentment. They were choosing to do the work, not being forced to. And they were having more success.

In a number of studies spanning the exercise and working world, inner drive led to more effort put forth, a higher degree of commitment, and better results. In a study of over one hundred British athletes, researchers found that inner drive not only resulted in better persistence on a demanding exercise test, but also helped in other aspects of toughness, such as seeing a challenge instead of a threat and using positive coping strategies instead of dissociative or disengagement strategies. Internally driven athletes experienced more positive emotions and an increased willingness to repeat the task after its completion.

The alternative, a motivation based on control, is fragile. It might appear powerful at the beginning, but it quickly fades. In the throes of despair between continuing and stopping, fear as the driver falters. In the aforementioned study, individuals motivated by external pressure were more likely to disengage, give up, and see the endeavor as a threat.

Not only were individuals with internal drive better at persisting; they had another secret weapon. They were better at reengaging. When it comes to striving for success, we often focus all of our energy on the persistence part of the equation. But persistence isn't always wanted or warranted. Think of the climber who is closing in on the summit of the mountain, yet fatigue is taking over her body, and her mind is losing its clarity thanks to oxygen deprivation. She has enough energy to make it to the top, but to survive, she needs to be able to make it back down the mountain. In an analysis of nearly one hundred years of deaths on Everest, of all the climbers who made it to the so-called death zone of over 8,000 meters altitude, only 10 percent died while making their way up the mountain, while 73 percent perished on their way back down. What's the tough decision in this case? To persist, sapping your energy before the riskiest part of your journey? Or abandon the goal and find a new one: making it down the mountain?

The ability to reengage, to shift your goal elsewhere, is a vital skill that tough individuals possess. After putting so much work into a singular pursuit, it can be nearly impossible to let it go. No one wants to fail, to come so close but turn away. But tough individuals possess the self-awareness to evaluate and weigh the contrasting pulls of the desire to continue to reach their goal and the reality of the demands they face, and the risks that come with those. All so that they can make the best possible choice. Instead of blindly persisting, if the right move is to "quit," they are able to reframe their goal, or find a new one and reengage with the activity. The climber's goal shifts from reaching the peak to returning safely down to her loved ones. Tough people don't live in a black-and-white world of success or failure. They are able to adjust and pour their ability to persist into a new worthwhile goal.

Not surprisingly, researchers found that the ability to reengage is linked not to control-based motivation but instead to a drive from within. Psychologists found that when motivation comes from the inside, individuals are more likely to adjust their actions to feedback that tells them that a goal is no longer attainable. They are more likely to listen to the voice that tells them to reengage with another pursuit. Sometimes, that means abandoning your goal and picking a new course. Other times, it means moving on from your A goal and toward your B goal.

If your goal starts to become unattainable, reengaging means switching to something still within your grasp. If you are struggling to put pen to paper on your novel, it's switching from trying to write a chapter to simply outlining your thoughts. Reengaging allows you to shift the target slightly, so that instead of slamming on the brakes, you find something that you can manage in the moment. It's moving from reaching the peak to getting down in one piece.

Inner drive brings clarity, allowing you to listen to your body so that you can make the right decision during difficult moments. Whether your goal is to improve your persistence or develop the ability to reengage once a goal seems to be lost, motivation coming from within appears to be the special sauce. Psychologists call it autonomous motivation.

Filling Your Basic Needs

In the 1970s, Edward Deci and colleagues gave what appeared to be wooden three-dimensional *Tetris* pieces to a group of twenty-four college students and told them to build a shape out of the blocks. For three days, the students who returned to the lab were shown a new shape and went to work on the blocks sitting in front of them. For half of the participants, day two brought a pleasant surprise. For each puzzle that they solved in the allotted time, they'd receive a monetary reward. Motivated by something beyond simply killing time for the fun of it, the participants upped their work ethic, spending longer stretches on solving the puzzle.

But when the participants returned for day three, the monetary reward was gone. It was back to making the shapes for the sake of making shapes. Not surprisingly, with an external incentive gone, their motivation dropped. Participants spent less time attempting to create new shapes and were more apt to quit playing with the blocks and simply sit there. The phenomenon we now know as extrinsic versus intrinsic motivation was born. Other scientists soon repeated the experiment in a range of different tasks and age groups. Before too long, researchers had replicated the effect in schoolchildren drawing and athletes playing sports. When some external reward or punishment was introduced, it shifted people's motivational habits.

Deci, along with another psychologist, Richard Ryan, had a radical idea. Their findings on what motivated people didn't apply just to doing homework or solving a problem, but to something far greater: their well-being. Deci and Ryan expanded their work on intrinsic motivation, declaring that we all have three basic and innate psychological needs. If we satisfy these needs, our well-being will improve, and we'll be self-motivated for growth and development. Self-determination theory (SDT) was born. It includes the need for autonomy, competence, and relatedness. Or stated another way, to feel in control, like you can make progress, and to belong.

Since the introduction of SDT, it has been investigated and applied to everything from parenting to teaching to substance abuse. And supporting Deci and Ryan's original hypothesis, needs satisfaction is linked to better health, ratings of well-being, and performance in a variety of domains. Autonomy, competence, and relatedness serve as our basic psychological needs. And fulfilling our basic needs helps not only with well-being, but also with the ability to persist.

For his PhD dissertation, John Mahoney combined the world of well-being and SDT with athletic prowess. Not only was there evidence connecting satisfying our three basic psychological needs to increased persistence, a variety of studies tied it to willingness to put forth effort, improved concentration, challenge seeking, and better coping with stress. When we have autonomy and support, we have higher self-esteem and better emotional intelligence. Satisfying our basic needs seemed to help with every attribute that underlies toughness.

In a series of studies, Mahoney and colleagues looked at cross-country runners and rowers, two individual sports where suffering and working through pain and fatigue are the norm. In over two hundred runners, satisfying the three basic needs was related to toughness and improved racing times. When Mahoney dug deeper into the data, he found that the social environment, primarily impacted by the coach, played a large role in whether athletes were able to meet these needs. If the atmosphere was supportive and fostered autonomy and a sense of belonging, then the athletes were tougher and performed better. As Mahoney concluded, toughness resulted from "coaching behaviors that promote psychological needs satisfaction."

When we satisfy our needs, we are allowed to fulfill our potential. Our drive comes from within, so fear and pressure no longer consume us. We feel like we belong so that even if we fail, we know that we'll still be loved and supported. We feel empowered, like we have control over the situation and can make an impact. Satisfying our basic needs is the fuel that allows us to put to work all of the tools we've developed to be tough. Without satisfying our basic needs, it doesn't matter how large our arsenal is to handle adversity.

TOUGHNESS MAXIM

When we satisfy our needs, we are allowed to fulfill our potential. Satisfying our basic needs is the fuel that allows us to put to work all of the tools we've developed to be tough.

If you're thinking, "Hmm, this runs contrary to what Bear Bryant, Bobby Knight, and my middle school gym teacher professed about creating tough teams," you'd be right. Being a demanding dictator? You've stripped your athlete of their autonomy, taking the decision away from them. Using fear and punishment or pushing people toward defaulting to surviving doesn't create intrinsic motivation; it creates the opposite. Yelling, screaming, getting in someone's face to push them forward? Same result: motivation via fear or pressure, which may seem to work in the short term but ultimately fails when it matters. Using control and power to force obedience? It falls by the wayside when it counts. Creating bonds through mutual suffering without true support? The old-school method of toughness runs contrary to just about every one of our basic needs. Could our middle school football coach really have been so wrong?

Organizational psychologist Erica Carleton teamed up with sports psychologist Mark Beauchamp to understand the impact of a coach's style on the players they lead. They selected fifty-seven NBA head coaches who were in the league between 2000 and 2006 to evaluate not only the immediate impact of a coach on a team but also the long-term impact. They scoured newspapers, magazines, and interviews, searching for stories and reports of a coach's leadership style. They dug for insight into each coach, addressing how they led their team and the methods they utilized before creating long scouting reports on each coach's style.

Once the coaching reports were finished, they handed them to trained psychologists to assess the leadership style of every coach. In particular, they were looking for their level of abusive leadership, which refers to when a coach utilizes ridicule or blame in an effort to motivate or teach those under their charge. Think: a coach telling players that they aren't competent or tough enough to perform. In other words, the behaviors and methods that we often ascribe to creating old-school toughness.

In evaluating almost seven hundred players' performance, those who played under a coach who utilized an abusive leadership style saw a clear drop in performance, as measured by a player efficiency score. But the effects weren't limited to the season in which they played under a coach who relied heavily on such tactics. The impact stretched to the player's entire career. According to their model, when a player experienced a highly abusive leadership style, the player's entire career trajectory was shifted a notch downward. Not only did their performance drop off, but the coach's style rubbed off on the players. Players who experienced an abusive leadership style had more technical fouls, an indicator of aggression, throughout the remainder of their careers. Keep in mind, these were players from the NBA, paid millions of dollars to win games. The researchers gave away their opinion of such coaching styles in the title of their paper, "Scarred for the Rest of My Career? Career-Long Effects of Abusive Leadership on Professional Athlete Aggression and Task Performance."

The leader—be it a CEO, manager, or coach—is the one who dictates how much autonomy individuals have. They are the ones who set the stage for belonging and making progress. Do they create a cult-style environment where individuals have little input and are instructed to follow? Are they allowed to take risks and explore their potential without extreme adverse consequences? Are people allowed to foster positive social support? Are they pushed toward seeing teammates as threats? The leader sets the tone, creating an environment that can either support or thwart athletes' basic needs. When those in charge choose the path of thwarting via control and power, subordinates' motivation shifts to pressure and fear. We see increases not only in aggression, but also in burnout, and we see decreases in performance and well-being. Controlling coaching and leading don't just harm performance; it harms the person.

On the other hand, leading via needs satisfaction helps create tougher, healthier, happier humans. As sports psychologist Laura Healy reported, "When athletes perceive their coaches to be more autonomy supportive, they report greater satisfaction of their basic psychological needs, and consequently strive for their goals with higher autonomous motives."

Contrary to our old-school expectations, developing toughness doesn't involve training camps from hell or exercise as punishment. It doesn't involve cruel, demanding bosses with little appreciation of the individual. It doesn't involve strict, one-way-communication parenting with little feedback from your child on their needs. As my good friend Brad outlined when explaining what he'd learned in his first few years of parenting, "At any moment, your child can go from fine to a fit. After a long day, it's easy to want to yell at them and tell them to grow up. But, especially with young kids, you have to realize that the emotions they're feeling are real. Even if they're inconsequential. It requires all the patience in the world, but I try to ask myself, how can I meet them at their level, explain it, and use it as a teachable moment. Yelling at them, exerting excessive control, all that does is teach them to be afraid of you. Why would I want to ingrain to only listen because they're afraid of me?" Fear is easy to instill. Trust is much harder. Instead of relying on fear and control, real toughness is linked to self-directed learning, feeling competent in your skills, being challenged but allowed to fail, and above all, feeling cared for by the team or organization.

In other words, toughness comes from the same building blocks that help create healthy, happy humans. Contrary to decades of ingrained ideology, toughness isn't developed through control or punishment; it's developed through care and support. If we take Deci and Ryan's self-determination theory and put a performance spin on it, then we're left with three key needs that leaders have to satisfy:

1. Being supported, not thwarted: having input, a voice, and a choice
2. The ability to make progress and to grow
3. Feeling connected to the team and mission; feeling like you belong

Can it be this simple? Does it actually work at the highest levels?

Support, Not Thwart

"I have to coach my team, I have not reached them for the last month. They're tired of my voice. I'm tired of my voice. It's been a long haul over these past few years." A coach who seems like he's at his wit's end, frustrated by his inability to connect with his team after what one could imagine was a series of disappointing seasons. We've all had those spells where we can't connect or get our message across to whomever we are guiding. We pull out all of our tricks, trying to teach utilizing every method we know, but are left with individuals who seem like they've lost hope. They tune us out and seemingly are going through the motions to "get through it."

That's why when you hear Steve Kerr, the coach of the Golden State Warriors, relay this message to reporters, it takes you aback. The Warriors weren't at the bottom of the NBA; they were at the top. In February 2018, when Kerr spoke these words, the Warriors were the defending champions and in the midst of one of the most dominant team runs in NBA history, with three championships and two finals losses in five years. At the time, Kerr's 2018 team was 44–13 and on its way to the third of those finals victories. Kerr wasn't speaking out of frustration. He was explaining why before win number 44 of the season, he turned over coaching duties to his players.

For the morning shootaround that preceded that evening's game, Kerr relinquished his duties to veteran small forward Andre Iguodala. When game time came, as the players huddled around during a timeout, it wasn't Kerr drawing up the plays and schemes; it was Iguodala, Draymond Green, and the rest of the team. They took control. After a sluggish first quarter, the Warriors found their coaching groove and ended up beating the Phoenix Suns 129–83. The stunt wasn't a gimmick by the innovative coach. As he explained to the press, Kerr felt like his team was losing focus. He wanted to send a message, as he reported after the game: "It's the players' team. It's their team, and they have to take ownership of it. . . . They determine their own fate, and I don't feel like we focused well at all the last month and it just seemed like the right thing to do."

In an autonomy-supportive environment, the leader acts as a guide, a person who is on the journey with the individual. The leader pokes, prods, nudges, and maybe even pushes individuals in certain directions, but the leader understands that they are there to help others reach their potential. That while they can direct and guide, it ultimately is up to the individual to take ownership of their actions.

In supportive environments, choice and ownership take center stage. When Kerr handed over the coaching reins, he was doing just that. Letting the team know that they were important and that he trusted them. Research shows that when leaders adopt such a model, their subordinates have better coping skills, are more self-confident, and are rated as more coachable.

On the other side of the coin is a leader who thwarts autonomy. They relish control and power. Dictating and directing with little input from the players. They rely on rewards, fear, punishment, and manipulation to maintain a sense of control. When researchers at Eastern Washington University compared coaches utilizing either servant (supportive) or power (thwarting) styles in sixty-four NCAA track teams, the athletes under the servant leader scored higher on measures of mental toughness and ran faster on the track. In the workplace, the story is much the same. In a recent study of over one thousand office workers, the strongest predictor of how well they dealt with the challenges of demanding work was whether they felt respected and valued by their managers. Their bosses simply showing they truly care led to increases in work engagement, loyalty, and resilience. Being a decent, caring human being is a performance and life enhancer.

When leading, you have to ask a simple question: What type

of motivation are you ingraining? Is it to be motivated via punishment or rewards, or is it to chase mastery? Are you dictating and controlling, sending the message that they should be motivated only when the boss is looking and tells them to do something, or are you handing the ball to your player, offering some guidance, and telling them to do their thing? Autonomy-supportive coaches and bosses work hard to foster choice, give their athletes and team members input, and allow for some control over their journey.

Can I Make Progress?
Objective sports are simple. You either ran faster, threw farther, lifted more, or jumped higher, or you didn't. Swimming, weight lifting, track and field, cycling, and similar sports are defined by inches, pounds, and seconds. There's no judge defining your performance; whether you are improving or getting worse is clear. When it's going well, it's magical. Athletes get on a roll, see their performance improve week after week, and transform into a beacon of confidence. However, when performance lags, objectivity turns from being helpful to being a hindrance. There's no rationalizing your individual performance because the team's still winning. There's no blaming a biased judge or referee. When performance starts to go south, athletes' self-belief seeps out of them. They start believing that breaking their previous personal best is an insurmountable task. Their motivation slips away, and the previously optimistic individual is replaced by a pessimistic version that can't see better performances on the horizon. When an athlete hits this low, it's nearly impossible to get them out of this rut. The ability to see a brighter future has disappeared. They are stuck.

Ensuring that we can see progress is essential for maintaining motivation. We need to see our story continuing, not that we've reached the final page of our book and have nowhere left to go. The same phenomenon occurs in every workplace. A previously engaged employee transforms into an apathetic mess when there is no path toward moving up in their career. We often think that bonuses and salary increases are what motivate individuals during their careers. The reality is there is no easier way to kill motivation than making future progress seem impossible. Once someone sees the goal as unattainable, complacency and apathy soon follow.

Being able to see yourself grow is a fundamental human need. As leaders, we need to create environments that allow people to see a brighter future that includes growth and mastery. That means providing pathways for moving up in the workplace and multiple different ways on which to judge success and growth. If it's only about a single metric or the bottom line, we're setting ourselves up for failure.

The final key to developing competence in the workplace is the ability to take risks and potentially fail. If you want to ensure that individuals stay where they are, then utilize fear of failure. If a person knows that if they fail at a project, then punishment, or potentially being let go, is the result, then you can be assured that they will not venture out of their comfort zone. They will take the necessary steps to ensure their survival. They won't take appropriate risks, try to innovate, or step outside of their comfort zone. In a fear-based environment, growth stagnates, even if there is a potential path to move up. Instead of leading by fear, workplaces that fill our basic needs have what's called psychological safety, or the ability to voice your thoughts and opinions without fear of punishment. Not to be confused with safe spaces, psychological safety is all about providing the security for people to take risks, to speak out, to be who they are. They can voice concerns to their superiors without fear of punishment. Throw out ideas without being shot down as wasting the company's time. When Google commissioned a two-year study on team performance, sitting atop their five characteristics of good teams was psychological safety. "Can we take risks on this team without feeling insecure or embarrassed?"

Cultivating an environment that allows for progress and competence has the following characteristics:

- A challenging but supportive environment
- The ability to take risks and voice your opinion without fear being the dominant motivator
- A path that shows the way for growth and improvement in your job or field

A Need to Belong
In 2010, Michael Kraus, Cassey Huang, and Dacher Keltner published a paper that made waves in the sporting world. For a number of games during the 2008–2009 NBA season, they tracked and coded nearly three hundred players' behaviors. No, they weren't looking at points, assists, blocks, or rebounds. Instead, they analyzed how often players showed acts of cooperation and trust. The fist bumps, high fives, short conversations, setting screens, and other actions, which showed either cooperation or lack of it with their teammates. Teams where the players high-fived, fist-bumped, and showed positive interactions with teammates more often had better performance across the season. The authors concluded that the high fives, chest bumps, and head slaps were tied to more team cooperation and better performance.

The study's takeaway wasn't what many in the sporting community interpreted it to be: go out and high-five your teammates.

The key wasn't the acts themselves. It's what they represented. Teams that are high in trust and belonging display more signals that they do just that, truly trust one another. Fist bumps communicate belonging. No different than a partner telling their significant other "I love you" to convey the status of their relationship, chest bumps send a clear message that "we are in this together. Great job, I've got your back." The goal isn't to increase fist bumps for better performance; it's to increase belonging.

Belonging is one of the foundational human needs. According to psychologist Scott Barry Kaufman, "When one feels belonging, one feels accepted and seen, and when one is deprived of belonging, one feels rejected and invisible." Humans are by and large social animals, dependent on the benefits of cooperation and connection for survival. If we remember back to chapter 6, where we discussed emotions as messengers, it should be no surprise that some of the most intense and unpleasant emotions are those tied to lack of connection. Loneliness, jealousy, shame, guilt, embarrassment, and social anxiety arise when we feel rejected from others. Feeling left out, rejected, is one of the most visceral emotions we experience. In fact, our brain interprets rejection in much the same way it does physical pain. There's a reason a broken heart can feel every bit as real as a broken arm. People have a deep need to feel valued, and when they don't, our most potent emotions let us know, begging us to do something about it.

We've all heard of the fight-or-flight stress system that helps us deal with threats and danger, but we possess another response that helps us build trust and belonging: the calm-and-connect system. When genuine connection occurs, feel-good opioids are released, which dampen down cortisol and other stress hormones, helping us transition out of a threat state. Another hormone, oxytocin, helps turn down the alarm system (amygdala) in our brain. When oxytocin is released, it leads to increased cooperation. One quirk of this hormone is that it only leads to increased connection when the person you share the experience with is deemed trustworthy. In other words, your brain has a system designed to help watch out for fake connection.

The calm-and-connect system is designed to take advantage of our social nature, creating connection so that we not only survive but thrive. It's no wonder that professional sports teams are trying to take advantage of this response. One of the cutting-edge strategies professional teams use to improve their recovery postgame isn't some proprietary protein shake or an expensive gadget. It's social interaction. So-called social recovery not only creates cohesion but also transitions from a stressful state (i.e., the game) to a recover-and-adapt state. Being around others helps us not only bond but recover. Connection is a secret weapon. But just like giving high fives, it can't be forced. Authentic connections happen during the in-between moments.

Gregg Popovich is the legendary coach of the San Antonio Spurs. When it comes to creating a team culture, he's the envy of just about every coach, CEO, or leader out there. His players rave about the bonds they still hold with their teammates years later. One player recently said, "I was friends with every single teammate I ever had in my [time] with the Spurs. That might sound far-fetched, but it's true." One key to such cohesion? Team meals.

Popovich dinner stories are almost apocryphal. Three-hour meals, with Popovich expertly choosing the wine and food. Tables arranged so that the players maximize interaction. Popovich eschews the NBA tradition of jetting out of the city straight after a game, and instead, the Spurs go straight to a meal and stay the night. No set agenda other than time together, and certainly no ropes courses or other contrived team-bonding activities.

In a 2003 study, researchers set out to understand why soldiers fight. Were they motivated by a sense of duty? In surveying a group of American soldiers who fought in the Iraq War, the deep emotional bond between soldiers came out on top. As the researchers dug deeper, they found that it wasn't the organized time training next to each other that mattered most; it was the in-between time. "What all of the research highlights is the importance of conversation during noncombat time—the hours of nothingness, the shared boredom—where bonds of trust, friendships, and group identity are built," write the researchers. Such moments allow us to get beyond the superficial, to realize that the person sitting next to me on the team bus, or in the cubicle down the hall, is a human being wrestling with the same issues that you and I are.

What Popovich and the military both discovered was the power of conversation to create connection. After the popularization of the Popovich story thanks to an ESPN article by Baxter Holmes, others tried to copy the popular coach's behavior. Teams across professional sports started implementing fancy postgame meals. Yet, when working with professional teams across a variety of sports, I kept hearing the same refrain: "We hosted team dinners, and no one showed up." Or "No one talked. It became a 'mandatory' activity that no one wanted to be at."

Like the fist-bump research, the magic isn't in the activity itself. The lesson isn't to have team dinners but to create a place for bonding and trust to develop. Popovich utilizes his power of conversation and passion for food and wine to create an environment where people want to be. He invited them into a world he truly cares about and lets his passion infect others. The time and care spent creating the perfect dinner makes people want to attend.

You can't force cohesion or unity. It doesn't come from trust falls, gimmicky bonding activities, or forced interaction. It comes from being real. From allowing people to lower their defenses and feel comfortable enough to be who they are. You can't force it. All you can do is create the space for it to happen. The magic wasn't in Popovich's team dinners. It was in his creating space for genuine interaction.

When we feel connected to those around us, we free ourselves up to perform. With a stable platform of support, we function from a place of growth and development instead of fear. This is as true in the workplace as it is on the courts. Belonging creates trust. Trust shifts the focus toward mastery for the greater good. Lack of connection and fear push us in the opposite direction, toward self-preservation, where everyone walks around just trying to make sure they survive. Belonging is expansive, freeing us up to play to win. Fear constricts, causing us to play to not lose.

Popovich was meticulous about the dinners he set up because he understood what psychology has discovered: your environment invites action. In the workplace, the same effect often occurs at the watercoolers and other liminal spaces. Coworkers have informal chats, exchanging ideas, connecting, and eventually coming to the realization that someone is more than just an accountant or manager. In the name of efficiency, there's often a misguided pull to overschedule and optimize every portion of the workday. Management tries to curtail time spent in the lounge, at lunch, or chitchatting in the hallway. It's a misuse of company time. But when we adopt such a mindset, we lose the in-between times. Instead of seeing a group discussing the *Game of Thrones* finale in the hallway as a waste of company time, see it as an opportunity to create cohesion. Progressive companies are adopting a science-backed trend, creating workplaces that try to increase moments of informal interaction. Instead of encouraging working through lunch or eating at your desk, create an environment that invites and promotes interaction. As a leader, it's your job to create space for genuine connection to occur. As I told a group of executives recently, if you walk around your office and no one is shooting the shit, you need to change the environment. Move furniture, set the example, engage with people in a shared passion. If you are at a company get-together or dinner and everyone is looking down at their phone, you've created the wrong environment.

In his book *The Culture Code*, author Dan Coyle outlines what he calls the vulnerability loop. Coyle makes the case that contrary to the way we traditionally think about it, we don't need trust before we can become vulnerable. The opposite is true. In order to trust, we first must be vulnerable. Being open and vulnerable sends an invitation to the person sitting across from you that you trust them. If that signal is reciprocated, that trust between the pair increases. We lower our defenses and guard, feeling open to being who we are. The more this cycle repeats, the stronger trust and cooperation become.

It's no wonder that when researchers looked at leadership and toughness in the athletic world, one of the best predictors of toughness was the relationship that an individual had with their peers and coaches. In a world dominated by social media and the appearance of connection, rather than actual relationships, the need for genuine connection is even more paramount. Look no further than the simple lesson that your grandparents likely passed on: eat family dinners together. A tradition going back centuries that's been shown to lower rates of depression and anxiety, substance abuse, eating disorders, and early pregnancy. When we spend time together with those we love and respect, good things follow. It's a virtuous cycle. The way to create a sense of belonging isn't some corporate retreat or a forced and artificial team bonding activity; it's in creating space for genuine, real connection.

Toughness = Filling Our Basic Needs

Before Edward Deci and Richard Ryan developed self-determination theory as a way to understand that our basic psychological needs drive motivation, Abraham Maslow created a hierarchy of needs, a theory that includes not only our psychological needs of safety, belonging, and self-esteem, but also our basic physiological needs of food, water, and sleep. Maslow's work set the stage for psychology to transition from looking at what is wrong with people to looking at what helps them grow and develop. Maslow wrote, "One can choose to go back toward safety or forward toward growth. Growth must be chosen again and again; fear must be overcome again and again."

Maslow's hierarchy is often pictured with self-actualization—a need for development, creativity, and growth—at the top of a pyramid. But that's not what Maslow intended. The famous pyramid that is associated with his hierarchy was created by someone else, not Maslow. In his 1970 journal, he wrote, "I realized I'd rather leave [self-actualization] behind me. Just too sloppy & too easily criticizable." To Maslow, self-actualization was too focused on the individual, satisfying our own needs in a somewhat selfish manner. The peak wasn't about the person, but something much greater. In Maslow's final rendering, self-transcendence is the highest rung. It's when we can reach beyond ourselves. As the phrase dictates, it's rising above individual concerns.

Maslow was convinced that many were capable of getting to self-transcendence, but they often got in their own way. Writing before his death that society often pushes us against it as "most industrialists will carefully conceal their idealism, their meta-motivations, and their transcendent experiences under a mask of toughness." For far too long, we've held on to this mask of toughness. We've fallen for tactics that do anything but satisfy our needs.

When we satisfy our basic psychological needs, we allow ourselves to reach their full potential, to utilize the tactics and strategies discussed throughout this book to work through challenging times. Our basic needs give us a stable platform to venture away from and come back to. We can handle fear and pressure because we know that if we do fail, we will still be loved and valued. We feel that we can make progress not just in our performance-related pursuit, but as better human beings. We don't get to this place through control; we get there through belonging, acceptance, and being allowed to be who we are. Who would have thought? The basic building blocks of being a healthy, functioning human are the same ones that enable us to handle tough situations. Let's stop getting in our own (or athletes', students', workers') way and work with our basic psychology and biology.

CHAPTER 11

Find Meaning in Discomfort

A forty-one-year-old man stood at the lectern. He had dark hair pulled back over his head. His round glasses and attire hinted at his importance and credentials, an MD who was also pursuing a PhD. It was a Saturday evening, 5 p.m., to be exact, in a lecture hall at an adult education college. For the next five Saturdays, the professor would deliver a series of hour-long lectures to whoever attended. The course description made clear the seriousness of the talks: "Suicide-forced annihilation, the world of the mentally ill, sexual education." Along with one final lecture that became known as the "Experimentum Crucis," a Latin phrase borrowed from Isaac Newton's breakthrough discoveries and used to describe an experiment that puts the nail in the coffin in one scientific theory and elevates the new theory to superiority.

The professor began the first lecture, "To speak about the meaning and value of life may seem more necessary today than ever; the question is only whether and how this is possible." While the topics of the lecture may have seemed to indicate to the listeners that they were sitting in an abnormal psychology class, the topics were a means to an end. Over five Saturdays, the professor outlined his theory on the keys to having a meaningful life. He railed against conventional wisdom, declaring that joy could not be pursued and that happiness "should not, must not, and can never be a goal, but only an outcome." Such pursuits did not bring meaning to our life. They did not fill our soul. Instead, if we wanted to be fulfilled, there were three ways to achieve this.

First, the act of doing. Creating—whether it be in an artistic pursuit or a labor of love—brought meaning to one's life. The second was in experiencing—nature, love, art, or anything that might create the sensation of awe and expand one's perspective of the world. You can imagine an audience sitting in a lecture hall being slightly thrown off by the joy and happiness declaration, but their minds were surely back in sync with the message of doing and experiencing to create meaning. The third key to fulfillment would catch most off guard, but this particular audience was likely expecting it: suffering.

The professor went on: "True suffering of an authentic fate is an achievement, and, indeed, is the highest possible achievement." No, the doctor, who was soon to have his PhD in philosophy, hadn't gone off the rails. To him, suffering wasn't to be sought out. But if we found ourselves there, meaning could be extracted. Suffering strips us of our vanities and allows us an opportunity to respond. To determine what our reaction would be to hardship and difficulty. To the man standing at the lectern, suffering wasn't just a way to develop meaning, but meaning was a way to get through suffering. To work through adversity, the suffering needed to be meaningful, and to the doctor, that was determined by "the individual, and only that individual," as he told his audience.

This wasn't the first time the doctor had given this lecture. He'd done so once before. He wasn't wearing a suit and tie. He wasn't standing in a lecture hall in front of a captive audience. It was a far different place. He was with 280 others, split into rows of 5. They weren't dressed in the business attire of the current audience. They were a ragged group, smaller in size and girth. Before the lecture began, small talk centered on the same topic: the soup that would be fed for dinner. The lecture wasn't vocal but in his mind. To distract from his current ordeal, to provide a brief escape for his mind, Viktor Frankl visualized being in the same lecture hall he was standing in now and began his lecture, titled "The Psychology of the Concentration Camp."

While few of us will face a challenge as horrifying as enduring a concentration camp, learning from those who went through unbelievable suffering provides lessons on navigating lesser problems. When the going gets tough, it's easy to lose meaning. When our job seems overwhelming, when we're on the brink of burnout, it's only natural to shrug our shoulders and ask "What's the point?" What those like Frankl, including researchers of the latest science, tell us is that purpose is the glue that holds us together, allowing us to rise above even the most harrowing of situations.

The Will to Continue

Willie is a six-year-old mutt, a mixture of Australian cattle dog and shepherd, that my wife found living in a tire when he was only a few months old. He's also got quite the personality, equally intelligent and mischievous. In his spare time, Willie sticks to his world of contrasts, having two loves: watching TV and running. The former is an amusing but also annoying habit he's picked up of pressing his large black nose to whatever screen has movement and sound coming out of it. Turn on a video of horses, and your screen will be destroyed. The latter act, running, is a more normal activity for a dog.

When Willie goes on a run, he makes it five miles in the chilly winter months, tugging at his leash the whole way. But whenever the heat and humidity of the summer come through, Willie's endurance evaporates. He trots through two miles with the same eagerness and excitement, but the heat quickly slows his pace from the usual seven-minute miles to a more pedestrian nine-minute-mile pace. Willie runs the same two-mile route every summer day; he knows the turns, taking them before we even signal to. He could pace it to perfection, but his pacing skills resemble the children discussed in chapter 3. Out hard, slow in the middle, before a spurt to the finish when we round the corner to the house. But there's something else that will cause Willie to abandon the fatigue-induced trot for a full-bore sprint: squirrels.

His pink tongue could be fully hanging out of his mouth, signaling that he's nearly had enough. He could have slowed to a panting-filled walk. But in the throes of fatigue, it all changes once he catches a glimpse of that small gray animal that torments his life. He springs up and begins to put his sixty-pound frame to work, pulling toward his mortal enemy. Willie found motivation, or dare I say meaning, to bring him back from the depths of despair.

In chapter 3, I discussed how children and adult endurance athletes typically decide to speed up or slow down. They use a heuristic, comparing actual versus expected effort for where they are at during the race. If they feel better than expected, they speed up. Worse? Time to slow down. But I intentionally left out one component: drive.

$$\text{Performance} = \text{Actual effort} / \text{Expected Effort} * \text{Drive}$$

Whether we call it drive, motivation, or purpose, the last component determines our bandwidth for how far into the depths of fatigue we can push. Contrary to common convention, when we are exhausted, we haven't completely depleted our reserves. Even the athlete who collapses to the ground after a race had more left in the tank. His or her muscles could still function. Just think, if our brain truly allowed us to go to zero, to empty, would that be smart or dangerous? Your body is trying to protect you, and it has a safety mechanism, an alarm that sounds to convince you to slow down, to stop, to make the pain and fatigue go away. But, according to the latest theories in exercise science, how close to zero we get is variable.

We are like the car with the countdown of how many miles we can drive before we hit zero. For the adventurous among us, you quickly learned that this gauge isn't entirely accurate. You may be able to drive ten, twenty, or even thirty miles past the point your car tells you that you have zero miles to go. There's leeway between when the vehicle manufacturers programmed the display to tell you that you were empty and when your gas tank is indeed empty. It's a safety mechanism designed to prevent people from tempting fate, pushing their car too far, and running out of gas on the side of the road. Our body utilizes the same type of process. Warning us—through the sensation of effort and fatigue—that we are at zero before we actually hit zero. There's always something in reserve. Drive determines how close to empty we can push before our body shuts us down. Before it flips from voluntary shutdown to initiating near catastrophe.

Our brain weighs that complicated gas mileage algorithm with how important the task is. Because we never truly reach zero in our body's energy reserves (that would mean catastrophic failure), we use the level of importance and the risk versus reward to determine how close to zero we can get. Is our life on the line? Is our child's life in jeopardy? Then, we might be able to perform superhuman feats like lifting a car off a trapped child's body. Is it a regular-season game, or is it game 7 of the finals? We might get a little extra juice in the latter case. According to the latest science of fatigue, your brain essentially tries to protect you from harming yourself, and it uses the perceived risks versus the potential rewards to fix where that governor is.

Having a strong purpose acts as a turbo boost. Whether that purpose comes from God, family, playing for your teammates, or a mission that holds deep importance, when our pursuits match with our purpose, we persist for much longer. Research shows that purpose and persistence are linked in the classroom, workplace, and athletic fields. When we have a purpose, we are able not only to endure and persist but also to provide a beacon that reminds us of what's important and to make the right decision at the right moment.

TOUGHNESS MAXIM

Purpose is the fuel that allows you to be tough.

From Fear to Despair to Apathy

Eleven months after being liberated, Viktor Frankl stood at the podium in a lecture hall in the outskirts of Vienna and delivered the talk he'd imagined while trapped in captivity. By 1946, he had published a book that changed psychotherapy, *The Doctor and the Soul*, and a book that the Library of Congress would dub one of "the ten most influential books in the US," *Man's Search for Meaning*. The latter he wrote in nine days. Each outlined the horrors of survival, but with an underlying current of hope. To go from concentration camp to producing works that are still relevant over a half century later is quite the feat. But as Frankl said, he'd been working on them for years.

Frankl entered the Theresienstadt concentration camp in 1942 with a book manuscript sewn into his jacket pocket. Like all possessions, the jacket was confiscated and with it the manuscript. Frankl had already developed his theory on meaning, and now he was unfortunately about to put it to the test. Over the next few years, Frankl experienced the unimaginable horror of living through the Holocaust. He lost his mother, father, and wife. Meaning willed him forward.

Frankl took his psychotherapist training and applied it to observing the conditions of the camp and its inhabitants. He witnessed how everyone, including himself, went through a series of phases. First, there was shock, of being stripped of all reminders of his former self, before momentarily looking for the quickest out. "Everyone in this situation flirts, if only for a moment, with the idea of running into the wire, committing suicide, using the usual method in the camp; contact with the high voltage barbed wire fence."

As the tragedy and uncertainty grow, individuals transition to a state of apathy, where the horror of their reality becomes almost normalized. Individuals who once reacted to death and despair with a standard emotional response now felt nothing. As he remarked, "Whereas in the first few days, the sheer abundance of experiences filled with ugliness—hateful in every sense—provoke feelings of horror, outrage, and disgust, these feelings eventually subside, and inner life as a whole is reduced to a minimum." The common remark to the death of another in the camp became a somber "Hmm." Not because it didn't matter, but because the inner world fell into hibernation. Feelings and emotions disappeared, replaced by a kind of nonresponsive protective state. And inner thoughts circled around the only thing that seemed to matter: food. In a study of other Holocaust survivors, researchers confirmed Frankl's experience, with two survivors named Lou and Esther reporting they felt "numb and living a day at a time. . . . Your brains don't work. First, you want to die. Then you want to survive."

According to Frankl, survival depended on your inner world. "Despite the cruelty visited on prisoners by the guards, the beatings, torture, and constant threat of death, there was one part of their lives that remained free: their own minds," Frankl reported in his book *Yes to Life*. The key to a free mind was seeing meaning in life, not just in the way we are used to thinking of it, as some greater purpose. But in every moment. That if we could find meaning in the minuscule portions of life, that something was greater than ourselves, then we would possess the will to survive or, if circumstances deemed otherwise, the peace to see meaning in our demise.

Freedom was the key to meaning. Freedom to choose how you saw and experienced every part of suffering. To Frankl, death contained meaning, just as life did. Freedom of your mind wasn't just about surviving; it was about being able to choose. That if you were going to die, you still had the ability to escape the place you were in, to go somewhere else, even if it was only in your mind. That even in death, "it was essential that we should die a death of our own and not the death that the SS had forced on us!" Frankl wrote that those who did not survive were not the weakest but among the strongest.

Finding meaning, in your circumstance, in your suffering, or as a purpose to guide and ground you, was the key to handling such utter atrocity. For Frankl, that meaning was simple: to return to his loved ones and to finish his life's work. As he relays in his autobiography, "I am convinced that I owe my survival, among other things, to my resolve to reconstruct that lost manuscript." Writing gave him meaning. A will to survive. That's why during labor camp, his mind drifted to a future lecture. And when someone provided a pencil and scraps of paper in 1945, it gave him the needed boost to bounce back from a bout of despair.

Frankl went into the Holocaust with ideas from his experience as a doctor. He came out experiencing death and despair and seeing firsthand how he and others coped with it. It led him to conclude, "The unshakeable belief in an unconditional meaning to life that, one way or another, makes life bearable. Because we have experienced the reality that human beings are truly prepared to starve if starvation has a purpose or meaning."

Much of the research in subsequent years confirms Frankl's experience. In a study on eighty-nine Holocaust survivors, Katarzyna Prot-Klinger found that they consistently highlighted the importance of support and belonging, having someone to return to, preserving a sliver of normalcy, and perhaps most importantly, luck. In a study of thirteen survivors, Roberta Greene found similar themes, including making choices, practicing inner control, making a conscious decision to live, celebrating life, and thinking positively. When asked how they survived, people focused in their answers on family and making meaning of their circumstances and the war. In another study on Holocaust survivors, sociologist Aaron Antonovsky concluded that having a sense of coherence or a way to make sense of the world is what mattered. Coherence was made up of three components: comprehensibility, manageability, and meaningfulness.

Meaning's impact extends beyond Holocaust survivors to sufferers of other traumas. While most of us are familiar with the debilitative phenomenon of post-traumatic stress disorder (PTSD), the positive cousin post-traumatic growth (PTG) is lesser-known. In research involving those who've suffered traumatic experiences, from natural disasters to seeing the death of friends and family to experiencing the horrors of being a prisoner of war, a surprisingly large percentage of survivors experience PTG.

You might think that it must be those who suffered a milder version of the trauma that were able to grow from it, but that's not the case. In a number of studies on former POWs from the Vietnam War, the longer a POW spent in captivity, the more physical injuries experienced, the higher the growth. With significant trauma, our worldview and the assumptions contained within them are shattered. Yet, this shattering of assumptions allows these individuals to make their way through misery and reach higher levels of personal strength and appreciation of life and what it brings. When their worlds are challenged, a search for meaning commences. In this search, they are able to reconstruct their inner narrative, acknowledging their personal strength in navigating catastrophe and redefining what matters in life. According to psychiatrist Adriana Feder and colleagues at the Mount Sinai School of Medicine, "Severe trauma triggers a search for meaning and a fundamental reconfiguration of a person's life goals."

Individuals who experience post-traumatic growth don't avoid the discomfort. They experience the same flood of emotions and rumination of their internal voices that everyone else experiences. They are able to sort through and explore discomfort. Instead of intrusive rumination that causes them to spiral, they alter their inner voice, using what psychologists call deliberate and constructive rumination. Akin to the calm conversation discussed earlier, deliberate rumination meant an internal dialogue that's focused on problem-solving, whereas self-talk was focused on reflecting on and dealing with the situation in a more controlled and nonjudgmental way. To switch our inner world from intrusive to deliberate, we need to have a sense of control and an ability to understand and regulate our emotions. When studying over 170 college students who had experienced suffering due to the death of loved ones, debilitating accidents, or violence in their home, George Mason University psychologist Todd Kashdan found that it was those who navigated their anxiety and trauma not via avoidance but through exploration who experienced the highest levels of growth.

When we explore instead of avoid, we are able to integrate the experience into our story. We're able to make meaning out of struggle, out of suffering. Meaning is the glue that holds our mind together, allowing us to both respond and recover. It stalls the jump from difficulty to complete despair, from fear and anxiety to full-blown freak-out. As Viktor Frankl said so many years ago when discussing the plight of those in concentration camps, "He retains a freedom, the human freedom to adapt to his fate, his environment, in one way or another—and indeed there was a 'one way or another.'" Meaning provides the freedom for us to choose.

TOUGHNESS MAXIM

When we explore instead of avoid, we are able to integrate the experience into our story. We're able to make meaning out of struggle, out of suffering. Meaning is the glue that holds our mind together, allowing us to both respond and recover.

The Nuance of Inner Strength

Step on a crack; break your mother's back. When I was a child, I believed this. No, not in the way most children did, with a slight tinge of fear of the silly rhyme coming true. But in the literal sense. I was also afraid that if I didn't touch the fronts and backs of doorknobs, someone in my family would die. Or that if I didn't complete my bedtime ritual of turning the alarm on and off seven times before getting in my bed on the same side, in the same way, every night, I might not wake up the next morning. Ever since I can remember, I have suffered from obsessive-compulsive disorder

(OCD). I didn't have a name for it. My blissfully unaware family called them "quirks." I didn't understand what they were, but every one of those experiences, and the danger that accompanied them, felt entirely real.

Growing up with OCD was bizarre but also normal. When you're young, you don't realize what you're thinking and processing, or even whether it's wrong. Having intrusive thoughts was normal. Experiencing thoughts of harm and death slowly became part of life. It was irrational, and I grew to know that, but it felt so real. It isn't until you move on from the naivete of the early years, when your executive function starts to come fully online and awareness kicks in, that you start to ask: What's wrong with me?

Asking that question as a teen was not easy. It was downright terrifying, coming to grips with the fact that others didn't have the same thoughts or fears I did. They didn't see harm every time they picked up a knife or were driving along the freeway. OCD was my burden. Something that I had to navigate largely by myself. Don't get me wrong: my parents were supportive, helpful whenever I literally thought I was going to die if I went to sleep. But they were naive. A Southern conservative family in the 1990s who had no awareness of mental health and still thought seeing a "shrink" was a sign of weakness or a label that would follow you for the rest of your life. In their minds, they were trying to protect me. Get through childhood without any long-lasting labels, be normal, and all would work out. They taught me avoidance. Push the bad thoughts away. Stay away from activities that trigger you. So as a child, navigating the inner world of intrusive thoughts was almost all on me.

While there are many different variations of OCD, the popular press gets the portrayal wrong. We see the compulsions, the obsessive cleaning, the rituals, and think that is the disorder. That's the end result. OCD is a condition of intrusive thoughts, mixed with a strong feeling and sensation that pushes us toward the compulsion. We complete the ritual as an act of soothing or coping with the thoughts and feelings. The key to fixing my obsessive need to touch the front and back of every doorknob wasn't to tell or train me to stop doing that. It was to disentangle the thought and feeling that nudged and sometimes shoved me toward that action.

In one form of OCD, thoughts and action become deeply intertwined. There's no space between them. This so-called thought-action fusion occurs when a person feels that merely thinking a terrible thought either can make it come true or is just as bad as if you went through with it. As we learned in chapter 7, intrusive thoughts occur fairly regularly in all of us. The difference is that in those who suffer from OCD, the intensity dial is turned up to 11. Sufferers genuinely believe that the momentary thought of turning their car into oncoming traffic or jumping off the balcony of their apartment means that they are actually going to do it. For most of us, that random message in our brain is brushed off, a misfire that we don't need to assign any meaning to. For OCD sufferers, it's real. The feeling of fear comes with it, and a behavior to cope soon follows.

It's not just the intertwining of thoughts and action that is the issue. Recent research shows that the brains of those who suffer from OCD have a hard time learning what's "safe" and what's not. Once something has been deemed threatening, it's as if it's carved in stone, while for the rest of the population, it's written in pencil. Researchers found that when they associated a green face with an electric shock, both OCD patients and a control group responded as if the face was a threat. However, when they took the shock away and showed the green face repeatedly, the control group quickly came to understand that it meant no danger here.

The OCD group couldn't let go of the lingering threat. The area in their brain related to processing safety signals—the prefrontal cortex—wasn't firing.

Now, think about what we've learned in relation to toughness. Thoughts and feelings interact, building upon each other, pushing us toward action. Our threat-detection center plays a large role, biasing us toward fight, flight, freeze, or whatever we might think is best. Notice the similarities? OCD sufferers have to deal with every element of our pattern of toughness, only with the system rigged against us. A deeply connected thought-feeling-urge pattern that misfires and misdirects.

Growing up, we saw OCD as a sort of defect, something we had to hide or overcome. Ignore it. Don't discuss it. It was something that signified that I was different or that there was something wrong with me. But as I've grown older, I've come to understand OCD as something different. It was reality, part of who I was. Something that I had to learn to accept and navigate. Fortunately, I suffer from a moderate version and am in no way discounting those who suffer from a more severe version. I was lucky enough to figure out ways to cope. Not by fighting or suppressing, but by gradually creating space between thought and action. Just enough where I could squeeze in a different way to cope instead of following through with a compulsion. I learned how to separate the thought from the feeling and urge. Rewiring my mind to recognize that a thought is a thought, and some are meaningless. I still suffer from OCD, the intrusive thoughts, and behavioral urges. It will always be part of who I am. But it taught me something: that those who society sees as weak often are the strongest on the inside.

***

While much of this book focuses on developing toughness in the individual and team, the greater issue is how we conceptualize toughness as a society. We prop up the brazen and boisterous, providing a platform for those who scream the loudest. We promote those who are brash and overconfident, even though their work and results don't merit the bravado. We support politicians who write books with titles promoting admirable values such as resilience, fortitude, healing during division, truth, and self-reliance, yet when it comes to acting in accordance with those proposed notions, our political elite ditch those oh-so-esteemed values. And then drive back and forth over them as long as they get their payday or accolades. We prop up the companies that create slick-looking ads promoting values of inclusion and diversity, all while the inner workings of those companies are littered with cultures of abuse, hostility, and harassment.

We've fallen for the appearance without the substance. We've chosen the glitzy Instagram-filter version of toughness. One that is staged, distorted, and dependent on choosing to live in a fantasy instead of embracing reality—acknowledging the struggles, the failure, the doubts, the insecurity. Real toughness is coming to terms with who we are and what we face, and making sense of and finding meaning in that struggle.

It's time to leave the old-school notion of toughness behind. The external and fake version might give us the momentary feeling of strength and power when we lead via fear and control. But it's temporary. It quickly fades away. And as I've shown over and over again, when it really matters, it fails. We've copied the early-twentieth-century football coach and drill sergeant model for far too long. We know where it leads. To something that looks like toughness on the outside, but upon closer inspection falls apart. It's time to turn away from a false idol and hold up a different kind of toughness. Society desperately needs us to emphasize the internal, not the external.

Real toughness is living in the nuance and complexity of the environment, bodies, and minds we inhabit. There is no one standard pathway to inner strength, no formula for making difficult decisions or dealing with the extremes of discomfort. Real toughness is about acceptance: of who you are, what you're going through, and the discomfort that often comes with it. It's living in that place of tension so that the needed space can be created to find the best path forward.

It's time to embrace a better, real kind of toughness. One that acknowledges our common humanity and slays the myth that old-school toughness promotes. My hope is that this book is a small step toward a major course correction, one that teaches our children that acting tough isn't the same thing as being tough. That being vulnerable and honest isn't a sign of weakness but a sign of strength. It's time to redefine toughness. It's more important now than ever before. Ditch the facade and the external. It's time to focus on true inner strength.

We're all capable of developing such inner strength, even those who might be labeled as weak or failures. Here's to embracing reality, being secure in who we are, embracing our feelings and emotions as information, fulfilling our basic human needs, and finding purpose and meaning in life to carry us through life's challenges. As biographer Joshua Wolf Shenk wrote about the man who had to figure out how to help the country he loved navigate its toughest period, "Lincoln, by whatever combination of habit and choice, took his own path. He did not pretend to be anything other than he was."

Be who you are. That's real toughness.

ACKNOWLEDGMENTS

First, I'd like to thank those who made this book possible. The many coaching clients, athletes, scientists, and performers whom I leaned on for stories, research, and guidance. Without your openness, and in many cases willingness to be guinea pigs, this book would not have been written. A particular shout-out to the many coaching clients who provided insight and showed what toughness was all about, both on the athletic fields and in the workplace. There are too many to name. Trust that your contribution is valued and appreciated. A special thanks to the individuals who allowed me to share their stories in this book: Matt Parmley, Drevan Anderson-Kaapa, Nate Pineda, Meredith Sorensen, Britani Gonzales, Brian Zuleger, Mark Freeman, Jim Denison, Joseph Mills, Phoebe Wright, Andy Stover, and Brian Barraza.

Next, I'd like to thank those who have paved the way to do the work. Thanks to my collaborative partner, Brad Stulberg. How much does your friendship mean to me? Well, as an introvert, I put up with on average about five phone calls a day from you. If that's not love, I don't know what is. Chris Douglas for steering the ship of The Growth Equation, which allows me to focus on the work. Jonathan Marcus and Danny Mackey for sticking by my side in the coaching world for over a decade now. To the coaches and colleagues who taught about toughness: Gerald Stewart, Mike Del Donno, Bob Duckworth, Tom Tellez, Theresa Fuqua, Leroy Burrell, Will Blackburn, and Kyle Tellez. To my friends and teammates: Chris Rainwater, Paulo Sosa, Frankie

Flores, Marcel Hewamudalige, Calum Neff, and many more. This book is the result of over twenty years of thought. It began out on the many runs and conversations we shared. I hope you see your imprint, philosophy, and thoughts littered throughout this book. To those in the random e-mail group who have offered support, wit, wisdom, and an inordinate amount of haikus along the way: Dave Epstein, Alex Hutchinson, Mike Joyner, Jonathan Wai, Amby Burfoot, and Christie Aschwanden. Thank you to the early readers who provided valuable feedback that made this book better: Chris Schrader, Howard Namkin, Peter Dobos, and Ben Wach. I hope you'll see the fruits of your labor. I listened, perhaps stubbornly, but I listened!

Publishing a book is a formidable task. One that a few short years ago I was clueless about. Thanks to the many who took a naive coach and turned him into a writer. Ted Weinstein for taking a chance on a couple of unknowns. Laurie Abkemeier, who is equal parts agent, advocate, editor, and sounding board. Without your help, I'm sure this book never would have seen the light of day. To the entire team at HarperOne, for believing in an idea, nurturing it, molding it, and turning it into something that hopefully makes a difference in the world. My publisher, Judith Curr, for taking a chance and turning this book into a reality. My editor, Anna Paustenbach. You believed in the vision of this book. You were a champion, a supporter, and an excellent editor who brought clarity and focus to the book. Amy Sather, Tanya Fox, and the rest of the team for coordinating, marketing, and turning a rough draft into the clear, hopefully insightful product you hold in your hand.

For a book on toughness, I must include a thank-you to the whistleblowers who have displayed more grit than I knew was imaginable. We thought running and racing were brutal. The actual test comes when you are put in a position to stand up and do the right thing, even when the easier path, to stay seated, is much more alluring. Few people stand up. You did. Thank you for doing so. Kara and Adam Goucher, Danny Mackey, Mary Cain, and the many others who chose the hard path and did the right thing.

And above all, to my wife, Hillary. Anytime I was struggling, stuck, or frustrated, you were there. You're the person I lean on, the person I look to, the person with unwavering support and confidence, even when I lack it. I'm a much better writer, thinker, and, most importantly, human being because of you. I love you.

NOTES

CHAPTER 1: FROM TOUGH COACHES, TOUGH PARENTS, AND TOUGH GUYS TO FINDING REAL INNER STRENGTH

4 *Looking back years later, Knight described*: B. Knight and B. Hammel, *Knight: My Story* (New York: Macmillan, 2002), 251.

4 *his coaching career at West Point*: L. Freedman, "Knight Focuses on Life Lessons," Cody Enterprise, May 1, 2017, https://www.codyenterprise.com/news/local/article_3aed785a-2eac-11e7-93cd-274ea4321321.html.

4 *"Now I'll fucking run your ass"*: E. Boehlert, "Why Bob Knight Should Bag It," Salon, April 3, 2000, https://www.salon.com/2000/04/03/knight_3.

5 *Baumrind defined responsiveness as*: D. Baumrind, "The Influence of Parenting Style on Adolescent Competence and Substance Use," *Journal of Early Adolescence* 11, no. 1 (1991): 56–95.

6 *In one study of over one thousand parents*: "The Fallacy of Tough Love: Queendom.com's Study Reveals That Authoritarian Parenting Can Do More Harm Than Good," PRWeb, August 6, 2013, https://www.prweb.com/releases/2013/8/prweb10996955.htm.

8 *"The stern approach is necessary"*: M. Hyman, *Until It Hurts: America's Obsession with Youth Sports and How It Harms Our Kids* (Boston: Beacon Press, 2009), 58.

9 *consistently shows quietly handle pain better*: L. J. Martin, E. L. Acland, C. Cho, W. Gandhi, et al., "Male-Specific Conditioned Pain Hypersensitivity in Mice and Humans," *Current Biology* 29, no. 2 (2019): 192–201.

9 *Instead of pulling the player from practice*: S. Almasy, "Maryland Football Player Who Died from Heat Stroke Needed Cold Immersion Therapy, Report Says," CNN, September 23, 2018. https://www.cnn.com/2018/09/22/us/maryland-jordan-mcnair-death-report/index.html.

9 *By the final sprint, video footage shows McNair*: The Diamondback, "Surveillance Footage of Maryland Football Player Jordan McNair's Final Workout," YouTube, December 20, 2018, https://www.youtube.com/watch?v=6EO_phwlAD0.

10 *A disease once primarily*: C. Aalborg, C. Rød-Larsen, I. Leiro, and W. Aasebø, "An Increase in the Number of Admitted Patients with Exercise-Induced Rhabdomyolysis," *Tidsskriftet Den Norske Legeforening* (2016).

10 *"Taking a cue from a head coach with a desire"*: B. D. Ridpath, "Oregon's Treatment of Athletes Is Unacceptable but Sadly It Is More Common Than People Realize," *Forbes*, January 20, 2017.

10 *Authoritarian parenting leads to*: N. Darling and L. Steinberg, "Parenting Style as Context: An Integrative Model," *Psychological Bulletin* 113, no. 3 (1993): 487–96; L. R. Williams, K. A. Degnan, K. E. Perez-Edgar, H. A. Henderson, et al., "Impact

of Behavioral Inhibition and Parenting Style on Internalizing and Externalizing Problems from Early Childhood through Adolescence," *Journal of Abnormal Child Psychology* (June 2009): 1063–75; and C. Jackson, L. Henriksen, and V. A. Foshee, "The Authoritative Parenting Index: Predicting Health Risk Behaviors among Children and Adolescents," *Health Education & Behavior* 25, no. 3 (1998): 319–37.

11 *On the athletic fields, it's linked to*: L. Scharneck, "The Mediating Effect of Self-Determined Motivation in Student-Athlete Perceptions of Coaching Behaviors and Its Effect on Grit and Mental Toughness," (diss., Illinois State University, 2017); and Y. Tabei, D. Fletcher, and K. Goodger, "The Relationship between Organizational Stressors and Athlete Burnout in Soccer Players," *Journal of Clinical Sport Psychology* 6, no. 2 (2012): 146–65.

11 *In one study of over 1,200 parents, authoritarian parenting*: "The Fallacy of Tough Love," PRWeb.

11 *those who grew up in an authoritarian environment*: O. Mayseless, M. Scharf, and M. Sholt, "From Authoritative Parenting Practices to an Authoritarian Context: Exploring the Person-Environment Fit," *Journal of Research on Adolescence* 13, no. 4 (2003): 427–56.

11 *"I need to work harder because"*: G. Kerr and A. Stirling, "Issues of Maltreatment in High Performance Athlete Development: Mental Toughness as a Threat to Athlete Welfare," in *The Handbook of Talent Identification and Development in Sport*. Routledge/Taylor and Francis. 409–20.

13 *According to one star player*: J. Corbett, "Pete Carroll Leads Seahawks with Enthusiasm, Toughness," USA Today, January 18, 2014, https://www.usatoday.com/story/sports/nfl/seahawks/2014/01/18/pete-carroll-seattle-seahawks-usc-49ers-super-bowl-lombardi/4637293.

14 *"Teaching guys how to feel confident"*: B. Schulze, "Pete Carroll: Mental Toughness Key to Seattle Seahawks Success," Bleacher Report, October 25, 2012, https://bleacherreport.com/articles/1384093-pete-carroll-mental-toughness-key-to-seattle-seahawks-success.

15 *"Despite the success of Wooden, Shula, Dungy, Stevens, and others"*: K. Reed, "It's Time to Bench Tyrannical Coaches," HuffPost, January 23, 2014, https://www.huffpost.com/entry/sports-coaches_b_4195220.

15 *researchers out of Eastern Washington*: M. Rieke, J. Hammermeister, and M. Chase, "Servant Leadership in Sport: A New Paradigm for Effective Coach Behavior," *International Journal of Sports Science & Coaching* 3, no. 2 (2008): 227–39.

20 *The American Thoracic Society cites "strong emotions"*: "What Are Vocal Cord Dysfunction (VCD) and Inspiratory Laryngeal Obstruction (ILO)?," American Thoracic Society, (n.d.), https://www.thoracic.org/patients/patient-resources/resources/vocal-cord-dysfunction.pdf.

20 *a shift in the nervous system activity*: J. G. Ayres and P. L. A. Gabbott, "Vocal Cord Dysfunction and Laryngeal Hyperresponsiveness: A Function of Altered Autonomic Balance?," *Thorax* 57, no. 4 (2002): 284–85.

CHAPTER 2: SINK OR SWIM: HOW WE TOOK THE WRONG LESSON FROM THE MILITARY

23 *As one student remarked of the time*: R. Coffey II, "The Bear Bryant Days at Aggieland, 1954–1957," The Association of Former Students, September 29, 2015, https://www.aggienetwork.com/news/140555/the-bear-bryant-days-at-aggieland-1954--1957.

23 *As senior quarterback Elwood Kettler recalled*: T. Badger, "'Junction Boys' Remembers Bear Bryant," *Plainview Herald*, December 11, 2002, https://www.myplainview.com/news/article/Junction-Boys-Remembers-Bear-Bryant-8937650.php.

23 *He was determined to harden his team*: J. Dent, "Ten Days in Hell with the Bear," ESPN, November 19, 2003, https://www.espn.com/classic/s/dent_junction_08/02/01.html.

23 *"The facilities were so sorry"*: T. Deas, "Gameday: Junction Revisited," Tuscaloosa News, September 13, 2013, https://www.tuscaloosanews.com/story/news/2013/09/14/gameday-junction-revisited/29910807007/.

24 *"It wasn't a football field, it wasn't any kind of field"*: R. Clark, "Survivors of A&M Coach 'Bear' Bryant's Grueling Training Camp Reunite in Junction on 60th Anniversary," *The Eagle*, August 15, 2014, https://www.theeagle.com/news/local/survivors-of-a-m-coach-bear-bryant-s-grueling-training/article_87a14b0e-eda4-5ade-8446-c7f23ff876f3.html.

24 *Practice was brutal, as Mickey Herskowitz reported*: P. Bryant and J. Underwood, *Bear: The Hard Life and Good Times of Alabama's Coach Bryant* (Triumph Books, 2007).

24 *"Sixth Player Quits Team at Texas A&M"*: "Sixth Player Quits Team at Texas A&M," *Washington Post and Times-Herald*, September 9, 1954, 29.

24 *"We don't care. First bus out"*: Dent, "Ten Days in Hell."

24 *Jim Dent's classic book*: Dent, "Ten Days in Hell."

24 *As Bob Easley, a fullback on the 1954 team, put it*: D. Barron, "Junction Boys Story Resonates after 60 Years," *Houston Chronicle*, August 16, 2014, https://www.houstonchronicle.com/sports/college-football/article/Junction-Boys-story-resonates-after-60-years-5693420.php.

25 *Only eight players who survived the camp*: M. Simonich, "'Junction Boys' Controversy: Key Figure in Bear Bryant Sports Biography Surfaces; Disputes Episode Alleging Coach Brutality," *Pittsburgh Post-Gazette*, December 4, 2002, http://old.post-gazette.com/ae/20021203junctionwebae2.asp.

25 *He stayed at home*: "1956 College Football All-America Team," Wikipedia, (n.d.), retrieved August 11, 2019, https://en.wikipedia.org/wiki/1956_College_Football_All-America_Team.

25 *Years later, Ed Dudley*: D. Andrews, "Dudley Recalls Days with Junction Boys," *Plainview Herald*, December 16, 2002, https://www.myplainview.com/news/article/Dudley-recalls-days-with-Junction-Boys-8861341.php.

26 *"That first year was brutal"*: Bryant and Underwood, *Bear*.

26 *The quitters included All-Southwest Conference players*: "Broussard Quits Ags;

Seventh to Leave," *Houston Chronicle*, September 8, 1954, B10, https://blog.chron.com/bayoucityhistory/files/2014/08/joined_document1.pdf.

26 *Foster "Tooter" Teague*: R. K. Wilcox, *Scream of Eagles: The Dramatic Account of the US Navy's Top Gun Fighter Pilots and How They Took Back the Skies over Vietnam* (New York: Simon and Schuster, 2005).

27 *for those who made it through camp*: R. Goldstein, "Jack Pardee, a Star at Texas A&M and an NFL Coach, Dies at 76," *New York Times*, April 2, 2013, https://www.nytimes.com/2013/04/03/sports/football/jack-pardee-texas-am-star-and-nfl-coach-dies-at-76.html.

27 *"Our instinct was survival"*: J. Dent, *The Junction Boys: How 10 Days in Hell with Bear Bryant Forged a Champion Team at Texas A&M* (New York: Macmillan, 1999).

27 *"it was the only way I knew how to do it"*: Badger, "'Junction Boys' Remembers Bear Bryant."

29 *Sixty-five percent of experienced soldiers*: C. A. Morgan III, G. Hazlett, S. Wang, E. G. Richardson Jr., et al., "Symptoms of Dissociation in Humans Experiencing Acute, Uncontrollable Stress: A Prospective Investigation," *American Journal of Psychiatry* 158, no. 8 (2001), 1239–47.

30 *having to endure a blaring speaker*: B. Webb, "What It's Like at the Training Camp Where Elite Soldiers Learn to Survive if They Are Captured and Tortured," Business Insider, December 19, 2015, https://www.businessinsider.com/sere-school-2015-12.

31 *The US Air Force's SERE*: Department of the Air Force, *Air Force Handbook: Survival Evasion Resistance Escape (SERE) Operations*, 2017, https://static.e-publishing.af.mil/production/1/af_a3/publication/afh10-644/afh10-644.pdf.

31 *the US Army is the largest employer of sports psychologists*: K. Weir, "A Growing Demand for Sport Psychologists," *Monitor on Psychology*, November 2018, https://www.apa.org/monitor/2018/11/cover-sports-psychologists.

32 *In 2014, the RAND Corporation*: S. Robson and T. Manacapilli, *Enhancing Performance under Stress: Stress Inoculation Training for Battlefield Airmen*, RAND Corporation, Project Air Force, 2014, https://apps.dtic.mil/dtic/tr/fulltext/u2/a605157.pdf.

32 *Navy SEALs recognized this distinction*: A. H. Taylor, S. Schatz, T. L. Marino-Carper, M. L. Carrizales, et al., "A Review of Military Predeployment Stress Tolerance Training," *Proceedings of the Human Factors and Ergonomics Society Annual Meeting* 55, no. 1 (2011): 2153–57.

33 *Comprehensive Soldier Fitness*: "Comprehensive Soldier Fitness," US Army Reserve, https://www.usar.army.mil/CSF/.

34 *the US Army enacted the Human Dimension Strategy*: US Army, *The Army Human Dimension Strategy*, 2015, https://caccapl.blob.core.usgovcloudapi.net/web/character-development-project/repository/human-dimension-strategy-2015.pdf.

38 *According to the latest scientific theories, the brain functions*: A. Peters, B. S. McEwen, and K. Friston, "Uncertainty and Stress: Why It Causes Diseases and How It Is Mastered by the Brain," *Progress in Neurobiology* 156 (2017): 164–88.

CHAPTER 3: ACCEPT WHAT YOU ARE CAPABLE OF

45 *Neuroscientist Jane Joseph*: J. B. MacKinnon, "The Strange Brain of the World's Greatest Solo Climber," *Nautilus*, June 28, 2018, http://nautil.us/issue/61/coordinates/the-strange-brain-of-the-worlds-greatest-solo-climber-rp.

45 *We call this a stress response*: "Understanding the Stress Response," Harvard Health, July 6, 2020, https://www.health.harvard.edu/staying-healthy/understanding-the-stress-response.

46 *"It's too scary"*: E. C. Vasarhelyi and J. Chin, *Free Solo*, National Geographic Documentary Films, 2018.

46 *Research consistently shows that tougher individuals*: A. Levy, A. Nicholls, and R. Polman, "Cognitive Appraisals in Sport: The Direct and Moderating Role of Mental Toughness," *International Journal of Applied Psychology* 2, no. 4 (2012): 71–76.

53 *the group of researchers summarized their findings*: A. P. Doran, G. B. Hoyt, M. D. Hiller Lauby, and C. A. Morgan III, "Survival, Evasion, Resistance, and Escape (SERE) Training," in *Military Psychology: Clinical and Operational Applications*, eds. C. H. Kennedy and E. A. Zillmer (Guilford Press, 2012), 306.

54 *In a series of studies in the Netherlands*: O. Stavrova, T. Pronk, and M. D. Kokkoris, "Choosing Goals That Express the True Self: A Novel Mechanism of the Effect of Self Control on Goal Attainment," *European Journal of Social Psychology* 49, no. 6 (2018): 1329–36.

56 *A group of French researchers*: Y. Daviaux, J.-B. Mignardot, C. Cornu, and T. Deschamps, "Effects of Total Sleep Deprivation on the Perception of Action Capabilities," *Experimental Brain Research* 232, no. 7 (2014): 2243–53.

56 *These findings led sports psychologist Thibault Deschamps*: J. K. Witt, S. A. Linkenauger, J. Z. Bakdash, J. S. Augustyn, et al., "The Long Road of Pain: Chronic Pain Increases Perceived Distance," *Experimental Brain Research* 192, no. 1 (2009): 145–48.

56 *hill-judging expertise*: M. Bhalla and D. R. Proffitt, "Visual-Motor Recalibration in Geographical Slant Perception," *Journal of Experimental Psychology: Human Perception and Performance* 25, no. 4 (1999): 1076–96.

57 *In 2018, a group of researchers out of University College London*: N. Garrett, A. M. González-Garzón, L. Foulkes, L. Levita, et al., "Updating Beliefs under Perceived Threat," *Journal of Neuroscience*, 38, no. 36 (2018): 7901–11.

57 *But under stress, as lead researcher Tali Sharot summarized*: T. Sharot, "Why Stressed Minds Are More Decisive," BBC Future, June 15, 2018, http://www.bbc.com/future/story/20180613-why-stressed-minds-are-better-at-processing-things.

58 *biasing us to feel more pain*: P. Goffaux, W. J. Redmond, P. Rainville, and S. Marchand, "Descending Analgesia: When the Spine Echoes What the Brain Expects," *Pain* 130, nos. 1–2 (2007): 137–43.

CHAPTER 4: TRUE CONFIDENCE IS QUIET; INSECURITY IS LOUD

60 *"Training gives me a feeling"*: J. Lovesey, "Straight Man in a Twisty Race," *Sports*

Illustrated Vault, June 1, 1964, https://vault.si.com/vault/1964/06/01/straight-man-in-a-twisty-race.

62 *And that's precisely what Hays and Bawden*: K. Hays, O. Thomas, I. Maynard, and M. Bawden, "The Role of Confidence in World-Class Sport Performance," *Journal of Sports Sciences* 27, no. 11 (2009): 1185–99.

63 *"I just couldn't concentrate"*: Hays et al., "The Role of Confidence."

67 *these correlational findings*: Will Storr, "'It Was Quasi-Religious': The Great Self-Esteem Con," *The Guardian*, June 3, 2017, https://www.theguardian.com/lifeandstyle/2017/jun/03/quasi-religious-great-self-esteem-con.

67 *titled Toward a State of Esteem*: California Task Force to Promote Self-Esteem and Personal and Social Responsibility, *Toward a State of Esteem: The Final Report of the California Task Force to Promote Self-Esteem and Personal and Social Responsibility*, California Department of Education, 1990, https://files.eric.ed.gov/fulltext/ED321170.pdf.

68 *"what the causes might be"*: Will Storr, "'It Was Quasi-Religious': The Great Self-Esteem Con," *The Guardian*, June 3, 2017, https://www.theguardian.com/lifeandstyle/2017/jun/03/quasi-religious-great-self-esteem-con.

69 *"Nope, it's just not true"*: J. Singal, "How the Self-Esteem Craze Took Over America: And Why the Hype Was Irresistible," *The Cut*, May 30, 2017, https://www.thecut.com/2017/05/self-esteem-grit-do-they-really-help.html.

69 *According to research, millennials*: J. M. Twenge and J. D. Foster, "Birth Cohort Increases in Narcissistic Personality Traits among American College Students, 1982–2009," *Social Psychological and Personality Science* 1, no. 1 (2010): 99–106.

69 *with a number of important factors*: U. K. Moksnes and G. A. Espnes, "Self-Esteem and Life Satisfaction in Adolescents: Gender and Age as Potential Moderators," *Quality of Life Research* 22, no. 10 (2013): 2921–28.

71 *"then your brain logically concludes"*: M. Freeman, *You Are Not a Rock: A Step-by-Step Guide to Better Mental Health (for Humans)* (New York: Penguin Books, 2018), 103.

72 *External regulation is defined*: M. S. Fortier, R. J. Vallerand, N. M. Brière, and P. J. Provencher, "Competitive and Recreational Sport Structures and Gender: A Test of Their Relationship with Sport Motivation," *International Journal of Sport Psychology* 26 (1995): 24–39.

73 *Men seem to be more susceptible to the fake variety*: C. Koerner, "Apparently a Whole Lot of Dudes Think They Could Take On Serena Williams in Tennis," BuzzFeed News, July 13, 2019, https://www.buzzfeednews.com/article/claudiakoerner/men-score-serena-williams-tennis.

73 *more than former president Donald Trump*: H. Britzky, "Everything Trump Says He Knows 'More about Than Anybody,'" Axios, January 5, 2019, https://www.axios.com/everything-trump-says-he-knows-more-about-than-anybody-b278b592-cff0-47dc-a75f-5767f42bcf1e.html.

74 *those who tend to shout the loudest*: V. Bohns, "Why Do We Shout When We Argue?

Lack of Confidence," *Wall Street Journal*, August 21, 2021, https://www.wsj.com/articles/why-do-we-shout-when-we-argue-lack-of-confidence-11629518461.

75 *call an action crisis*: V. Brandstätter and J. Schüler, "Action Crisis and Cost-Benefit Thinking: A Cognitive Analysis of a Goal-Disengagement Phase," *Journal of Experimental Social Psychology* 49, no. 3 (2013): 543–53.

75 *Ming Ming Chiu*: C. Anzalone, "Overconfidence among Teenage Students Can Stunt Crucial Reading Skills," University at Buffalo, 2009, http://www.buffalo.edu/news/releases/2009/07/10284.html.

76 *As Ilona Jerabek, president of the company*: "The Validity of the 'Fake-It-Till-You-Make-It' Philosophy," PRWeb, April 13, 2019, https://www.prweb.com/releases/the_validity_of_the_fake_it_till_you_make_it_philosophy/prweb16239903.htm.

82 *"is to build your fundamentals"*: J. MacMullan, "Rise above It or Drown: How Elite NBA Athletes Handle Pressure," ESPN, May 29. 2019, https://www.espn.com/nba/story/_/id/26802987/rise-drown-how-elite-nba-athletes-handle-pressure.

82 *Social psychologist Heidi Wayment*: S. B. Kaufman, "The Pressing Need for Everyone to Quiet Their Egos," *Scientific American* Blog Network, May 21, 2018, https://blogs.scientificamerican.com/beautiful-minds/the-pressing-need-for-everyone-to-quiet-their-egos.

84 *"thriving in adverse circumstances"*: J. Meggs, "Examining the Cognitive, Physiological and Behavioural Correlates of Mental Toughness," Teesside University, 2013, https://research.tees.ac.uk/en/studentTheses/examining-the-cognitive-physiological-and-behavioural-correlates-.

CHAPTER 5: KNOW WHEN TO HOLD 'EM AND WHEN TO FOLD 'EM

85 *Solomon's lab believed with fear conditioning*: S. F. Maier and M. E. Seligman, "Learned Helplessness at Fifty: Insights from Neuroscience," *Psychological Review* 123, no. 4 (2016): 349–67.

85 *Martin Seligman and Steven Maier*: Maier and Seligman, "Learned Helplessness at Fifty."

86 *They had lost the ability to try*: M. E. Seligman and S. F. Maier, "Failure to Escape Traumatic Shock," *Journal of Experimental Psychology* 74, no. 1 (1967): 1–9.

88 *In a letter from 1610, William Strachey writes*: "First-Hand Accounts," Virtual Jamestown, (n.d.), http://www.virtualjamestown.org/fhaccounts_desc.html#vaco.

88 *While another colonist reported*: K. O. Kupperman, "Apathy and Death in Early Jamestown," *The Journal of American History* 66, no. 1 (1979): 24–40.

89 *In 1620, colonist George Thorpe reported*: Kupperman, "Apathy and Death in Early Jamestown."

90 *More recently, individuals who were lost at sea*: H. Massey, J. Leach, M. Davis, and V. Vertongen, "Lost at Sea: The Medicine, Physiology and Psychology of Prolonged Immersion," *Diving and Hyperbaric Medicine* 47, no. 4 (2017): 239–47.

90 *rats conditioned for learned helplessness*: C. P. Richter, "On the Phenomenon of Sudden Death in Animals and Man," in *Psychopathology*, eds. C. F. Reed, I. E. Alexander, and S. S. Tomkins (Cambridge, MA: Harvard University Press, 2013), 234–42.

90 *According to a Senate report, the CIA*: "The Senate Committee's Report on the CIA's Use of Torture," *New York Times*, December 9, 2014.

91 *[Give-up-itis] is the clinical expression*: J. Leach, "'Give-Up-Itis' Revisited: Neuropathology of Extremis," *Medical Hypotheses* 120 (2018): 14–21.

92 *After receiving his MD from Emory University in 1962, Peter Bourne*: P. G. Bourne, R. M. Rose, and J. W. Mason, "17-OHCS Levels in Combat: Special Forces 'A' Team under Threat of Attack," *Archives of General Psychiatry* 19, no. 2 (1968): 135–40.

93 *When we lack control*: A. M. Bollini, E. F. Walker, S. Hamann, and L. Kestler, "The Influence of Perceived Control and Locus of Control on the Cortisol and Subjective Responses to Stress," *Biological Psychology* 67, no. 3 (2004): 245–60.

93 *When researchers peered into the brains*: T. V. Salomons, R. Nusslock, A. Detloff, T. Johnstone, et al., "Neural Emotion Regulation Circuitry Underlying Anxiolytic Effects of Perceived Control over Pain," *Journal of Cognitive Neuroscience* 27, no. 2 (2015): 222–33.

94 *When we believe we have influence over an outcome*: J. P. Bhanji, E. S. Kim, and M. R. Delgado, "Perceived Control Alters the Effect of Acute Stress on Persistence," *Journal of Experimental Psychology General* 145, no. 3 (2016): 356–65.

95 *According to self-determination theory*: "Self-Determination Theory," *Wikipedia*, (n.d.), retrieved January 5, 2020, https://en.wikipedia.org/wiki/Self-determination_theory.

96 *Albert Bandura's seminal theory of self-efficacy*: M. P. Carey and A. D. Forsyth, "Teaching Tip Sheet: Self-Efficacy," American Psychological Association, (n.d.), https://www.apa.org/pi/aids/resources/education/self-efficacy.

96 *The striatum, an area linked to reward processing*: L A. Leotti, S. S. Iyengar, and K. N. Ochsner, "Born to Choose: The Origins and Value of the Need for Control," *Trends in Cognitive Sciences* 14, no. 10 (2010): 457–63.

96 *When we are given that same reward based on luck or chance*: J. O'Doherty, P. Dayan, J. Schultz, R. Deichmann, et al., "Dissociable Roles of Ventral and Dorsal Striatum in Instrumental Conditioning," *Science* 304, no. 5669 (2004): 452–54.

96 *Giving nursing home residents more autonomy and choice*: E. J. Langer and J. Rodin, "The Effects of Choice and Enhanced Personal Responsibility for the Aged: A Field Experiment in an Institutional Setting," *Journal of Personality and Social Psychology* 34, no. 2 (1976): 191–98.

96 *While in the workplace*: S. Saragih, "The Effects of Job Autonomy on Work Outcomes: Self Efficacy as an Intervening Variable," *International Research Journal of Business Studies* 4, no. 3 (2011): 203–15.

97 *As Maier told the American Psychological Association*: S. F. Dingfelder, "Old Problem, New Tools," *Monitor on Psychology*, October 2009, https://www.apa.org/monitor/2009/10/helplessness.

99 *"We are born to choose"*: Leotti, Iyengar, and Ochsner, "Born to Choose."

101 *While the* New York Times *reported in 2015 that*: J. Kantor and D. Streitfeld, "Inside Amazon: Wrestling Big Ideas in a Bruising Workplace," *New York Times*, August 15, 2015, https://www.nytimes.com/2015/08/16/technology/inside-amazon-wrestling-big-ideas-in-a-bruising-workplace.html.

102 *Denison and Mills continued*: J. Denison and J. P. Mills, "Planning for Distance Running: Coaching with Foucault," *Sports Coaching Review* 3, no. 1 (2014): 1–16.

103 *A study of over two hundred men and women*: J. W. Mahoney, D. F. Gucciardi, N. Ntoumanis, and C. J. Mallett, "Mental Toughness in Sport: Motivational Antecedents and Associations with Performance and Psychological Health," *Journal of Sport and Exercise Psychology* 36, no. 3 (2014): 281–92.

105 *It's why research shows in dieting*: R. C. do Vale, R. Pieters, and M. Zeelenberg, "The Benefits of Behaving Badly on Occasion: Successful Regulation by Planned Hedonic Deviations," *Journal of Consumer Psychology* 26, no. 1 (2016): 17–28.

112 *a task is completely meaningless*: D. I. Cordova and M. R. Lepper, "Intrinsic Motivation and the Process of Learning: Beneficial Effects of Contextualization, Personalization, and Choice," *Journal of Educational Psychology* 88 (1996): 715–30.

CHAPTER 6: YOUR EMOTIONS ARE MESSENGERS, NOT DICTATORS

115 *"So it is a lie"*: E. Young, "The Only Emotions I Can Feel Are Anger and Fear," *Mosaic*, May 28, 2018, https://mosaicscience.com/story/life-without-emotions-alexithymia-interoception.

118 *Nerve fibers run throughout*: A. D. Craig, "Interoception: The Sense of the Physiological Condition of the Body," *Current Opinion in Neurobiology* 13, no. 4 (August 2003): 500–505.

118 *this sensory network has a name—interoception*: L. F. Barrett and W. K. Simmons, "Interoceptive Predictions in the Brain," *Nature Reviews Neuroscience* 16, no. 7 (2015): 419–29.

118 *The interoceptive system is active*: A. D. Craig, "How Do You Feel—Now? The Anterior Insula and Human Awareness," *Nature Reviews Neuroscience* 10, no. 1 (2009): 59–70.

119 *In 1896, Wilhelm Wundt*: R. B. Zajonc, "Feeling and Thinking: Preferences Need No Inferences," *American Psychologist* 35, no. 2 (1980): 151–75.

120 *over 70 percent of students had experienced*: S. Pareek, "Phantom Vibration Syndrome: An Emerging Phenomenon," *Asian Journal of Nursing Education and Research* 7, no. 4 (2017): 596–97.

120 *the more dependent you are on your phone*: D. J. Kruger and J. M. Djerf, "Bad Vibrations? Cell Phone Dependency Predicts Phantom Communication Experiences," *Computers in Human Behavior* 70 (2017): 360–64.

123 *They turned anxiety into excitement*: J. Strack dos Santos Gonçalves, P. Lopes, F. Esteves, and P. Fernández-Berrocal, "Must We Suffer to Succeed?: When Anxiety

Boosts Motivation and Performance," *Journal of Individual Differences* 38, no. 2 (April 2017): 113–24.

123 *In a series of studies, the patients*: L. Young, A. Bechara, D. Tranel, H. Damasio, et al., "Damage to Ventromedial Prefrontal Cortex Impairs Judgment of Harmful Intent," *Neuron* 65, no. 6 (2010), 845–51.

125 *Kate Winslet, the actress who played Rose in the film*: J. Denham, "'I Think He Could Have Fit on That Bit of Door': Kate Winslet Says Titanic Blunder Led to Leo DiCaprio's Movie Death," *Independent*, February 3, 2016, https://www.independent.ie/entertainment/movies/movie-news/i-think-he-could-have-fit-on-that-bit-of-door-kate-winslet-says-titanic-blunder-led-to-leo-dicaprios-movie-death-34419861.html.

125 *Even Cameron inserted himself into the fray*: R. Keegan, "James Cameron on *Titanic*'s Legacy and the Impact of a Fox Studio Sale," *Vanity Fair*, November 26, 2017, https://www.vanityfair.com/hollywood/2017/11/james-cameron-titanic-20th-anniversary-avatar-terminator-fox-studios-sale.

126 *When participants saw justified violence*: "Justified and Unjustified Movie Violence Evoke Different Brain Responses," The Annenberg Public Policy Center of the University of Pennsylvania, December 10, 2019, https://www.annenbergpublicpolicycenter.org/justified-movie-violence-unjustified-evoke-different-brain-responses-study-finds.

127 *In 2001, Harvard psychologist Joshua Greene*: J. D. Greene, R. B. Sommerville, L. E. Nystrom, J. M. Darley, et al., "An fMRI Investigation of Emotional Engagement in Moral Judgment," *Science* 293, no. 5537 (2001): 2105–8.

128 *nearly 17 percent of teenagers have cut themselves*: D. Gillies, M. A. Christou, A. C. Dixon, O. J. Featherston, et al., "Prevalence and Characteristics of Self-Harm in Adolescents: Meta-Analyses of Community-Based Studies 1990–2015," *Journal of the American Academy of Child and Adolescent Psychiatry* 57, no. 10 (October 2018): 733–41.

129 *When Young and colleagues compared*: Young, Hayley A., Dr, Jason Davies, and David Benton. 2019. "Non-suicidal Self-injury Is Associated with Multidimensional Deficits in Interoception: Evidence from Three Studies." PsyArXiv. April 24. doi:10.31234/osf.io/2azer.

130 *Impaired interoceptive awareness*: S. S. Khalsa, R. Adolphs, O. G. Cameron, H. D. Critchley, et al., "Interoception and Mental Health: A Roadmap," *Biological Psychiatry: Cognitive Neuroscience and Neuroimaging* 3, no. 6 (2018): 501–13.

131 *Research shows that tougher athletes*: A. Diaz, *The Relationship between Body Awareness and Mental Toughness in Collegiate Athletes*, doctoral dissertation, The Chicago School of Professional Psychology, 2013.

131 *A study out of the University of California San Diego*: L. Haase, J. L. Stewart, B. Youssef, and A. C. May, "When the Brain Does Not Adequately Feel the Body: Links between Low Resilience and Interoception," *Biological Psychology* 113 (2016): 37–45.

131 *psychologists found that stock traders*: N. Kandasamy, S. N. Garfinkel, L. Page, B.

Hardy, et al., "Interoceptive Ability Predicts Survival on a London Trading Floor," *Scientific Reports* 6, no. 1 (2016): 1–7; and Khalsa, Adolphs, Cameron, Critchley, et al., "Interoception and Mental Health."

132 *"current body states to predict future body states"*: Haase, Stewart, Youssef, May, et al., "When the Brain Does Not."

132 *Better interoceptive skills are correlated*: H. D. Critchley and S. N. Garfinkel, "Interoception and Emotion," *Current Opinion in Psychology* 17 (2017): 7–14.

133 *When researchers out of UCLA*: M. D. Lieberman, N. I. Eisenberger, M. J. Crockett, S. M. Tom, et al., "Affect Labeling Disrupts Amygdala Activity in Response to Affective Stimuli," *Psychological Science* 18, no. 5 (2007): 421–28.

137 *researchers out of Spain*: Strack dos Santos Gonçalves, Lopes, Esteves, and Fernández-Berrocal, "Must We Suffer to Succeed?"

138 *A systematic review*: I. Pedraza Ramirez, "Systematic Review of the Evidence of Interoceptive Awareness in Performers," 2016, https://jyx.jyu.fi/handle/123456789/51424.

CHAPTER 7: OWN THE VOICE IN YOUR HEAD

140 *"Water, Captain. Please"*: S. Callahan, *Adrift: Seventy-Six Days Lost at Sea* (Boston: Houghton Mifflin Harcourt, 2002), 195.

141 *As Callahan reported*: Callahan, *Adrift*, 56.

143 *researchers have identified at least seven*: D. T. Kenrick and V. Griskevicius, *The Rational Animal: How Evolution Made Us Smarter Than We Think* (New York: Basic Books, 2013).

144 *In one scene, Disgust laments*: P. Docter and R. del Carmen, *Inside Out*, Walt Disney Studios Motion Pictures, 2015.

145 *watched part of a scary movie*: J. K. Maner, D. T. Kenrick, D. V. Becker, T. E. Robertson, et al., "Functional Projection: How Fundamental Social Motives Can Bias Interpersonal Perception," *Journal of Personality and Social Psychology* 88, no. 1 (2005): 63–78.

146 *Over 94 percent of people*: M. Gannon, "Most People Have Unwanted, Worrying Thoughts," LiveScience, April 8, 2014, https://www.livescience.com/44687-most-people-have-unwanted-thoughts.html.

148 *five different voices*: M. M. Puchalska-Wasyl, "Self-Talk: Conversation with Oneself? On the Types of Internal Interlocutors." *The Journal of Psychology* 149, no. 5 (2015): 443–460.

148 *In another study, different forms*: P. K. McGuire, D. A. Silbersweig, R. M. Murray, A. S. David, et al., "Functional Anatomy of Inner Speech and Auditory Verbal Imagery," *Psychological Medicine* 26, no. 1 (1996): 29–38.

149 *In the book* The Voices Within: C. Fernyhough, *The Voices Within: The History and Science of How We Talk to Ourselves* (New York: Basic Books, 2016), 106.

150 *"the dominance of inner speech"*: Fernyhough, *The Voices Within*.

153 *modern research has largely validated Vygotsky's theory*: B. Alderson-Day and C. Fernyhough, "Inner Speech: Development, Cognitive Functions, Phenomenology, and Neurobiology," *Psychological Bulletin* 141, no. 5 (2015): 931–65.

154 *Research from clinical psychologist Steven Hayes*: S. C. Hayes, I. Rosenfarb, E. Wulfert, E. D. Munt, et al., "Self-Reinforcement Effects: An Artifact of Social Standard Setting?," *Journal of Applied Behavior Analysis* 18, no. 3 (1985): 201–14.

155 *Psychologists Pamela Highlen and Bonnie Bennett*: P. S. Highlen and B. B. Bennett, "Elite Divers and Wrestlers: A Comparison between Open- and Closed-Skill Athletes," *Journal of Sport and Exercise Psychology* 5, no. 4 (1983): 390–409.

155 *researchers out of the University of Waterloo*: J. V. Wood, W. Q. Perunovic, and J. W. Lee, "Positive Self-Statements: Power for Some, Peril for Others," *Psychological Science* 20, no. 7 (July 2009): 860–66.

156 *When sports psychologist Judy Van Raalte*: J. L. Van Raalte, B. W. Brewer, P. M. Rivera, and A. J. Petitpas, "The Relationship between Observable Self-Talk and Competitive Junior Tennis Players' Match Performances," *Journal of Sport and Exercise Psychology* 16 (1994): 400–15.

156 *In 2016, in a collaborative study*: R. E. White, E. O. Prager, C. Schaefer, E. Kross, et al., "The 'Batman Effect': Improving Perseverance in Young Children," *Child Development* 88, no. 5 (2017): 1563–71.

158 *Psychologists have used the same paradigm*: S. Rudert, R. Greifeneder, and K. Williams (eds.), *Current Directions in Ostracism, Social Exclusion and Rejection Research* (London: Routledge, 2019).

159 *psychologist Ethan Kross*: J. S. Moser, A. Dougherty, W. I. Mattson, B. Katz, et al., "Third-Person Self-Talk Facilitates Emotion Regulation without Engaging Cognitive Control: Converging Evidence from ERP and fMRI," *Scientific Reports* 7, no. 1 (2017): 1–9.

160 *Alain de Botton suggested*: A. de Botton, "Self-Love," The School of Life Articles, September 24, 2020, https://www.theschooloflife.com/thebookoflife/self-love.

CHAPTER 8: KEEP YOUR MIND STEADY

164 *Antoine Lutz and his colleagues*: A. Lutz, D. R. McFarlin, D. M. Perlman, T. V. Salomons, et al., "Altered Anterior Insula Activation During Anticipation and Experience of Painful Stimuli in Expert Meditators," *NeuroImage* 64 (2013): 538–46.

166 *yoga masters were able to*: R. Kakigi, H. Nakata, K. Inui, N. Hiroe, et al., "Intracerebral Pain Processing in a Yoga Master Who Claims Not to Feel Pain during Meditation," *European Journal of Pain* 9, no. 5 (2005): 581–89.

166 *For the everyday person*: T. R. Kral, B. S. Schuyler, J. A. Mumford, M. A. Rosenkranz, et al., "Impact of Short- and Long-Term Mindfulness Meditation Training on Amygdala Reactivity to Emotional Stimuli," *NeuroImage* 181 (2018): 301–13.

166 *associated with a hyperactive amygdala*: T. T. Yang, A. N. Simmons, S. C. Matthews, S. F. Tapert, et al., "Adolescents with Major Depression Demonstrate Increased

Amygdala Activation," *Journal of the American Academy of Child and Adolescent Psychiatry* 49, no. 1 (2010): 42–51.

166 *A recent study out of Yale*: A. L. Gold, R. A. Morey, and G. McCarthy, "Amygdala–Prefrontal Cortex Functional Connectivity during Threat-Induced Anxiety and Goal Distraction," *Biological Psychiatry* 77, no. 4 (2015): 394–403.

167 *According to the latest scientific research*: Kral, Schuyler, Mumford, Rosenkranz, et al., "Impact of Short- and Long-Term Mindfulness Meditation Training."

167 *Burnout is epidemic*: S. Ju, "16 Employee Burnout Statistics HR Leaders Should Know," Spring Health, December 14, 2020, https://www.springhealth.com/16-statistics-employee-burnout.

167 *weaker connection to their PFC*: A. Michel, "Burnout and the Brain," Association for Psychological Science, January 29, 2016, https://www.psychologicalscience.org/observer/burnout-and-the-brain.

169 *"less emotional elaboration of physiological cues"*: M. A. Rosenkranz, A. Lutz, D. M. Perlman, D. R. Bachhuber, et al., "Reduced Stress and Inflammatory Responsiveness in Experienced Meditators Compared to a Matched Healthy Control Group," *Psychoneuroendocrinology* 68 (2016): 117–25.

169 *In a group of over one hundred research subjects*: B. S. Schuyler, T. R. Kral, J. Jacquart, C. A. Burghy, et al., "Temporal Dynamics of Emotional Responding: Amygdala Recovery Predicts Emotional Traits," *Social Cognitive and Affective Neuroscience* 9, no. 2 (2012): 176–81.

169 *affective inertia*: S. Pichon, E. A. Miendlarzewska, H. Eryilmaz, and P. Vuilleumier, "Cumulative Activation during Positive and Negative Events and State Anxiety Predicts Subsequent Inertia of Amygdala Reactivity," *Social Cognitive and Affective Neuroscience* 10, no. 2 (2015): 180–90.

170 *"Whereas the long-term meditator is simply responding"*: E. Klein, "How the Brains of Master Meditators Change," Vox, May 30, 2019, https://www.vox.com/podcasts/2019/5/30/18644106/richard-davidson-ezra-klein-show.

171 *as little as four days of mindfulness*: F. Zeidan, K. T. Martucci, R. A. Kraft, N. S. Gordon, et al., "Brain Mechanisms Supporting the Modulation of Pain by Mindfulness Meditation," *Journal of Neuroscience* 31, no. 14 (2011): 5540–48.

172 *"pause between stimulus and response"*: R. May, *The Courage to Create* (New York: W. W. Norton & Company, 1975), 100.

175 *Research shows that when we practice*: Lutz, McFarlin, Perlman, Salomons, et al., "Altered Anterior Insula Activation."

177 *dark-room meditation*: "Dark Retreats," Samyama, (n.d.), https://samyama.com/dark-retreats.

177 *can improve our perceptual awareness*: R. Schuling, N. van Herpen, R. de Nooij, W. T. de Groot, et al., "Silent into Nature: Factors Enabling Improvement in a Mindful Walking Retreat in Nature of People with Psychological Symptoms," *Ecopsychology* 10, no. 2 (2018): 77–86.

177 *psychologist Timothy Wilson*: T. D. Wilson, D. A. Reinhard, E. C. Westgate, D. T. Gilbert, et al., "Just Think: The Challenges of the Disengaged Mind," *Science* 345, no. 6192 (2014): 75–77.

179 *When we train our ability*: C. N. Ortner, S. J. Kilner, and P. D. Zelazo, "Mindfulness Meditation and Reduced Emotional Interference on a Cognitive Task," *Motivation and Emotion* 31, no. 4 (2007): 271–83.

180 *Clinical psychologists have utilized*: S. S. Khalsa, R. Adolphs, O. G. Cameron, H. D. Critchley, et al., "Interoception and Mental Health: A Roadmap," *Biological Psychiatry: Cognitive Neuroscience and Neuroimaging* 3, no. 6 (2018): 501–13.

180 *In one study, psychologist Regina Lapate*: R. C. Lapate, B. Rokers, D. P. M. Tromp, N. S. Orfali, et al., "Awareness of Emotional Stimuli Determines the Behavioral Consequences of Amygdala Activation and Amygdala-Prefrontal Connectivity," *Scientific Reports* 6, no. 1 (2016): 1–16.

188 *"Now I can discuss it with you, and it's no big deal"*: S. Gregory, "Lolo's No Choke," *Time*, July 9, 2021, 30–38.

191 *"It is evenness of mind"*: B. Bodhi, "Toward a Threshold of Understanding," Access to Insight, 1998, https://www.accesstoinsight.org/lib/authors/bodhi/bps-essay_30.html.

191 *Hindu scripture proclaims to*: A. B. S. Prabhupada, *Bhagavad-Gita as It Is* (Los Angeles: Bhaktivedanta Book Trust, 1972), 104.

192 *"you will both avoid much misery"*: J. Wesley, *"Wesley's Notes on the Bible,"* Christian Classics Ethereal Library, 1765, https://www.ccel.org/ccel/wesley/notes.i.iv.xxii.html.

CHAPTER 9: TURN THE DIAL SO YOU DON'T SPIRAL

196 *A group of researchers out of Aarhus University in Denmark*: M. Clasen, M. Andersen, and U. Schjoedt, "Adrenaline Junkies and White-Knucklers: A Quantitative Study of Fear Management in Haunted House Visitors," *Poetics* 73 (2019): 61–71.

198 *Following the 1972 Olympics*: P. Milvy (ed.), *The Marathon: Physiological, Medical, Epidemiological, and Psychological Studies*, vol. 301 (New York Academy of Sciences, 1977).

198 *Frank Shorter was there*: S. Farrell, "The 1975 Elite Runners Study: How Are Elite Distance Runners Different from the Rest of Us?," The Cooper Institute, May 29, 2019, https://www.cooperinstitute.org/2019/05/29/the-1975-elite-runners-study-how-are-elite-distance-runners-different-from-the-rest-of-us.

199 *William Morgan and Michael Pollock*: W. P. Morgan and M. L. Pollock, "Psychologic Characterization of the Elite Distance Runner," *Annals of the New York Academy of Sciences* 301, no. 1 (1977): 382–403.

200 *In one study, 40 percent*: F. Dehais, M. Causse, F. Vachon, N. Régis, et al., "Failure to Detect Critical Auditory Alerts in the Cockpit: Evidence for Inattentional Deafness," *Human Factors* 56, no. 4 (2014): 631–44.

202 *They wanted to see if they could prime*: R. S. Friedman, A. Fishbach, J. Förster, and L. Werth, "Attentional Priming Effects on Creativity," *Creativity Research Journal* 15, nos. 2–3 (2003): 277–86.

203 *believe that rumination occurs when*: H. DeJong, E. Fox, and A. Stein, "Does Rumination Mediate the Relationship between Attentional Control and Symptoms of Depression?," *Journal of Behavior Therapy and Experimental Psychiatry* 63 (2019): 28–35.

204 *The broaden-and-build theory*: M. A. Cohn, B. L. Fredrickson, S. L. Brown, J. A. Mikels, et al., "Happiness Unpacked: Positive Emotions Increase Life Satisfaction by Building Resilience," *Emotion* 9, no. 3 (2009): 361–68.

205 *neuroscientists Noa Herz, Moshe Bar, and Shira Baror*: N. Herz, S. Baror, and M. Bar, "Overarching States of Mind," *Trends in Cognitive Sciences* 24, no. 3 (2020): 184–99.

210 *going for a walk out in nature*: D. K. Brown, J. L. Barton, and V. F. Gladwell, "Viewing Nature Scenes Positively Affects Recovery of Autonomic Function Following Acute-Mental Stress," *Environmental Science & Technology* 47, no. 11 (2013): 5562–69; and K. J. Williams, K. E. Lee, T. Hartig, L. D. Sargent, et al., "Conceptualising Creativity Benefits of Nature Experience: Attention Restoration and Mind Wandering as Complementary Processes," *Journal of Environmental Psychology* 59 (2018): 36–45.

211 *"break your heart and fill it up again"*: K. Arnold, *Running Home: A Memoir* (New York: Random House, 2019).

214 *"advanced the field of emotion regulation"*: G. Sheppes, S. Scheibe, G. Suri, and J. J. Gross, "Emotion-Regulation Choice," *Psychological Science* 22, no. 11 (2011): 1391–96.

214 *But researchers have found that rumination helps*: L. J. Altamirano, A. Miyake, and A. J. Whitmer, "When Mental Inflexibility Facilitates Executive Control: Beneficial Side Effects of Ruminative Tendencies on Goal Maintenance," *Psychological Science* 21, no. 10 (2010): 1377–82.

214 *So-called deliberate rumination*: K. Taku, A. Cann, R. G. Tedeschi, and L. G. Calhoun, "Intrusive versus Deliberate Rumination in Posttraumatic Growth across US and Japanese Samples," *Anxiety, Stress, and Coping* 22, no. 2 (2009): 129–36.

214 *researchers found that suppression works well*: G. A. Bonanno and D. Keltner, "Facial Expressions of Emotion and the Course of Conjugal Bereavement," *Journal of Abnormal Psychology* 106, no. 1 (1997): 126–37.

216 *The successful grievers*: S. Gupta and G. A. Bonanno, "Complicated Grief and Deficits in Emotional Expressive Flexibility," *Journal of Abnormal Psychology* 120, no. 3 (2011): 635–43.

216 *Flexible coping is tied*: I. R. Galatzer-Levy, C. L. Burton, and G. A. Bonanno, "Coping Flexibility, Potentially Traumatic Life Events, and Resilience: A Prospective Study of College Student Adjustment," *Journal of Social and Clinical Psychology* 31, no. 6 (2012): 542–67.

216 *Sheppes and Gross offered the following summary*: G. Sheppes, S. Scheibe, G. Suri, P.

Radu, et al., "Emotion Regulation Choice: A Conceptual Framework and Supporting Evidence," *Journal of Experimental Psychology: General* 143, no. 1 (2014): 163–81.

217 *At the 1988 Olympic Trials*: J. M. Silva and M. I. Appelbaum, "Association-Dissociation Patterns of United States Olympic Marathon Trial Contestants," *Cognitive Therapy and Research* 13, no. 2 (1989): 185–92.

219 *In tracking how children develop*: W. S. Grolnick, L. J. Bridges, and J. P. Connell, "Emotion Regulation in Two-Year-Olds: Strategies and Emotional Expression in Four Contexts," *Child Development* 67, no. 3 (1996): 928–41.

223 *"Your whole being is involved"*: J. Geirland, "Go with the Flow," *Wired*, September 1, 1996, https://www.wired.com/1996/09/czik.

225 *"Prolonged effortless concentration of attention"*: Y. Dormashev, "Flow Experience Explained on the Grounds of an Activity Approach to Attention," in *Effortless Attention: A New Perspective in the Cognitive Science of Attention and Action*, ed. B. Bruya (Cambridge, MA: MIT Press, 2010), 306.

225 *In another study, Swann and colleagues*: C. Swann, A. Moran, and D. Piggott, "Defining Elite Athletes: Issues in the Study of Expert Performance in Sport Psychology," *Psychology of Sport and Exercise* 16 (2015): 3–14.

226 *psychologist Scott Barry Kaufman stated*: S. B. Kaufman, *Transcend: The New Science of Self-Actualization* (New York: Penguin Random House, 2021).

CHAPTER 10: BUILD THE FOUNDATION TO DO HARD THINGS

230 *four key elements that controlling leaders utilize*: K. J. Bartholomew, N. Ntoumanis, and C. Thøgersen-Ntoumani, "The Controlling Interpersonal Style in a Coaching Context: Development and Initial Validation of a Psychometric Scale," *Journal of Sport and Exercise Psychology* 32, no. 2 (2010): 193–216.

231 *As the opening sentence of a* Forbes *article*: Coursey, David. "Steve Jobs Was a Jerk, You Shouldn't Be." *Forbes Magazine*, May 16, 2012. https://www.forbes.com/sites/davidcoursey/2011/10/12/steve-jobs-was-a-jerk-you-shouldnt-be/?sh=23998e0c4045.

231 *Implying that without incentive to work*: Z. Budryk, "'Mnuchin: It 'Wouldn't Be Fair to Use Taxpayer Dollars to Pay More People to Sit Home,'" The Hill, July 26, 2020, https://thehill.com/homenews/coronavirus-report/509062-mnuchin-it-wouldnt-be-fair-to-use-taxpayer-dollars-to-pay-more.

232 *covering over 120 years of study*: T. A. Judge, R. F. Piccolo, N. P. Podsakoff, J. C. Shaw, et al., "The Relationship between Pay and Job Satisfaction: A Meta-Analysis of the Literature," *Journal of Vocational Behavior* 77, no. 2 (2010): 157–67.

232 *a Gallup study of over 1.4 million employees*: J. Harter and N. Blacksmith, "Majority of American Workers Not Engaged in Their Jobs," Gallup, October 28, 2011, http://www.gallup.com/poll/150383/majority-american-workers-not-engaged-jobs.aspx.

232 *have a carrot dangled in front of them*: Y. J. Cho and J. L. Perry, "Intrinsic Motivation and Employee Attitudes: Role of Managerial Trustworthiness, Goal Directedness, and Extrinsic Reward Expectancy," *Review of Public Personnel Administration* 32, no. 4 (2012): 382–406.

232 *In a study of over one hundred British athletes*: N. Ntoumanis, L. C. Healy, C. Sedikides, J. Duda, et al., "When the Going Gets Tough: The 'Why' of Goal Striving Matters," *Journal of Personality* 82, no. 3 (2014): 225–36.

233 *In an analysis of nearly one hundred years*: P. G. Firth, H. Zheng, J. S. Windsor, A. I. Sutherland, et al., "Mortality on Mount Everest, 1921–2006: Descriptive Study," *BMJ* 337 (2008).

234 *Not surprisingly, researchers found*: N. Ntoumanis and C. Sedikides, "Holding On to the Goal or Letting It Go and Moving On?: A Tripartite Model of Goal Striving," *Current Directions in Psychological Science* 27, no. 5 (2018): 363–68.

235 *In the 1970s, Edward Deci and colleagues*: E. L. Deci, "Effects of Externally Mediated Rewards on Intrinsic Motivation," *Journal of Personality and Social Psychology* 18, no. 1 (1971): 105–15.

236 *supporting Deci and Ryan's original hypothesis*: M. Vansteenkiste, J. Simons, W. Lens, K. M. Sheldon, et al., "Motivating Learning, Performance, and Persistence: The Synergistic Effects of Intrinsic Goal Contents and Autonomy-Supportive Contexts," *Journal of Personality and Social Psychology* 87, no. 2 (2004): 246–60.

236 *For his PhD dissertation, John Mahoney*: J. W. Mahoney, D. F. Gucciardi, N. Ntoumanis, and C. J. Mallett, "Mental Toughness in Sport: Motivational Antecedents and Associations with Performance and Psychological Health," *Journal of Sport and Exercise Psychology* 36, no. 3 (2014): 281–92.

239 *"Scarred for the Rest of My Career"*: E. L. Carleton, J. Barling, A. M. Christie, M. Trivisonno, et al., "Scarred for the Rest of My Career? Career-Long Effects of Abusive Leadership on Professional Athlete Aggression and Task Performance," *Journal of Sport and Exercise Psychology* 38, no. 4 (2016): 409–22.

239 *those in charge choose the path of thwarting*: L. C. Healy, N. Ntoumanis, J. Veldhuijzen van Zanten, and N. Paine, "Goal Striving and Well-Being in Sport: The Role of Contextual and Personal Motivation," *Journal of Sport and Exercise Psychology* 36, no. 5 (2014): 446–59.

239 *As sports psychologist Laura Healy reported*: Healy, Ntoumanis, Veldhuijzen van Zanten, and Paine, "Goal Striving and Well-Being in Sport."

240 *"I have to coach my team"*: D. Kurtenbach, "Kurtenbach: Steve Kerr Turned In His Best Coaching Performance of the Year . . . by Not Coaching," *The Mercury News*, February 13, 2018, https://www.mercurynews.com/2018/02/13/warriors-v-suns-highlights-coaching-staff-andre-iguodala-draymond-green-timeouts-drawing-plays-golden-state-phoenx-roster-standings.

241 *"It's the players' team"*: A. Gilberg, "Steve Kerr Lets Andre Iguodala, Draymond Green Coach the Warriors during 129–83 Blowout Win over Suns," *Daily News*, February 13, 2018, https://www.nydailynews.com/sports/basketball/kerr-lets-iguodala-draymond-coach-warriors-blowout-win-article-1.3817084.

242 *Research shows that when leaders adopt*: J. W. Mahoney, D. F. Gucciardi, S. Gordon, and N. Ntoumanis, "Psychological Needs Support Training for Coaches: An Avenue for Nurturing Mental Toughness," in *Applied Sport and Exercise Psychology: Practitioner Case Studies*, eds. S. T. Cotterill, N. Weston, and G. Breslin

(London: Wiley, 2016), 193–213; and J. Mahoney, N. Ntoumanis, C. Mallett, and D. Gucciardi, "The Motivational Antecedents of the Development of Mental Toughness: A Self-Determination Theory Perspective," *International Review of Sport and Exercise Psychology* 7, no. 1 (2014): 184–97.

242 *When researchers at Eastern Washington University*: C. S. Hammer, "Mental Toughness, Servant Leadership, and the Collegiate Distance Runner," master's thesis, Eastern Washington University, 2012, https://dc.ewu.edu/theses/32.

242 *In a recent study of over one thousand office workers*: D. LaGree, B. Houston, M. Duffy, and H. Shin, "The Effect of Respect: Respectful Communication at Work Drives Resiliency, Engagement, and Job Satisfaction among Early Career Employees," *International Journal of Business Communication*, May 20, 2021, https://journals.sagepub.com/doi/abs/10.1177/23294884211016529.

245 *"without feeling insecure or embarrassed"*: J. Rozovsky, "The Five Keys to a Successful Google Team," Google re:Work, November 17, 2015, https://rework.withgoogle.com/blog/five-keys-to-a-successful-google-team.

245 *Michael Kraus, Cassey Huang, and Dacher Keltner*: M. W. Kraus, C. Huang, and D. Keltner, "Tactile Communication, Cooperation, and Performance: An Ethological Study of the NBA," *Emotion* 10, no. 5 (2010): 745–49.

246 *According to psychologist Scott Barry Kaufman*: S. B. Kaufman, *Transcend: The New Science of Self-Actualization* (New York: Penguin Random House, 2021), 38.

246 *Loneliness, jealousy, shame, guilt*: M. R. Leary, "Emotional Responses to Interpersonal Rejection," *Dialogues in Clinical Neuroscience* 17, no. 4 (2015): 435–41.

247 *His players rave*: B. Holmes, "Michelin Restaurants and Fabulous Wines: Inside the Secret Team Dinners That Have Built the Spurs' Dynasty," ESPN, July 25, 2020, http://www.espn.com/nba/story/_/id/26524600/secret-team-dinners-built-spurs-dynasty.

248 *In a 2003 study, researchers set out to understand*: L. Wong, "Why They Fight: Combat Motivation in the Iraq War," Strategic Studies Institute, 2003.

249 *Popovich was meticulous*: J. J. Waring and S. Bishop, "'Water Cooler' Learning: Knowledge Sharing at the Clinical 'Backstage' and Its Contribution to Patient Safety," *Journal of Health Organization and Management* 24, no. 4 (2010): 325–42.

250 *It's no wonder that when researchers*: C. Li, R. Martindale, and Y. Sun, "Relationships between Talent Development Environments and Mental Toughness: The Role of Basic Psychological Need Satisfaction," *Journal of Sports Sciences* 37, no. 18 (2019): 2057–65.

250 *eat family dinners together*: J. Anderson, "Harvard EdCast: The Benefit of Family Mealtime," Harvard Graduate School of Education, April 1, 2020, https://www.gse.harvard.edu/news/20/04/harvard-edcast-benefit-family-mealtime.

251 *Maslow wrote*: S. B. Kaufman, "Choose Growth," *Scientific American* Blog Network, April 7, 2020, https://blogs.scientificamerican.com/beautiful-minds/choose-growth/.

251 *In his 1970 journal*: M. Davis, "Maslow's Forgotten Pinnacle: Self-Transcendence,"

Big Think, August 9, 2019, https://bigthink.com/personal-growth/maslow-self-transcendence?rebelltitem=3#rebelltitem3.

251 *that society often pushes us against it*: A. Maslow, "Theory Z," W. P. Laughlin Foundation, https://atpweb.org/jtparchive/trps-01-69-02-031.pdf.

CHAPTER 11: FIND MEANING IN DISCOMFORT

253 *He railed against conventional wisdom*: V. E. Frankl, *Yes to Life: In Spite of Everything* (Boston: Random House, 2020), 32.

258 *Research shows that purpose and persistence*: R. A. Voorhees, "Toward Building Models of Community College Persistence: A Logit Analysis," *Research in Higher Education* 26, no. 2 (1987): 115–29; and A. M. Grant, "Does Intrinsic Motivation Fuel the Prosocial Fire?: Motivational Synergy in Predicting Persistence, Performance, and Productivity," *Journal of Applied Psychology* 93, no. 1 (2008): 48.

259 *"contact with the high voltage barbed wire fence"*: Frankl, *Yes to Life*, 88.

259 *In a study of other Holocaust survivors*: R. R. Greene, "Holocaust Survivors: A Study in Resilience," *Journal of Gerontological Social Work* 37, no. 1 (2002): 3–18.

260 *That even in death*: Frankl, *Yes to Life*, 97.

260 *As he relays in his autobiography*: V. E. Frankl, *Recollections: An Autobiography* (Cambridge, MA: Basic Books, 2008), 98.

261 *"Because we have experienced the reality"*: Frankl, *Yes to Life*, 37.

261 *In a study on eighty-nine Holocaust survivors*: K. Prot, "Strength of Holocaust Survivors," *Journal of Loss and Trauma: International Perspectives on Stress & Coping* 17, no. 2 (2012): 173–86.

261 *to live, celebrating life, and thinking positively*: Greene, "Holocaust Survivors."

261 *comprehensibility, manageability, and meaningfulness*: A. Antonovsky, *Unraveling the Mystery of Health: How People Manage Stress and Stay Well* (San Francisco: Jossey-Bass, 1987).

261 *the positive cousin post-traumatic growth*: A. Feder, S. M. Southwick, R. R. Goetz, Y. Wang, et al., "Posttraumatic Growth in Former Vietnam Prisoners of War," *Psychiatry: Interpersonal and Biological Processes* 71, no. 4 (2008): 359–70.

262 *According to psychiatrist Adriana Feder*: Feder, Southwick, Goetz, Wang, et al., "Posttraumatic Growth in Former Vietnam Prisoners of War."

262 *deliberate and constructive rumination*: C. J. Park and S.-K. Yoo, "Meaning in Life and Its Relationships with Intrinsic Religiosity, Deliberate Rumination, and Emotional Regulation," *Asian Journal of Social Psychology* 19, no. 4 (2016): 325–35; and M. Brooks, N. Graham-Kevan, M. Lowe, and S. Robinson, "Rumination, Event Centrality, and Perceived Control as Predictors of Post-Traumatic Growth and Distress: The Cognitive Growth and Stress Model," *British Journal of Clinical Psychology* 56, no. 3 (2017): 286–302.

262 *When studying over 170 college students*: T. B. Kashdan and J. Q. Kane, "Post-

Traumatic Distress and the Presence of Post-Traumatic Growth and Meaning in Life: Experiential Avoidance as a Moderator," *Personality and Individual Differences* 50, no. 1 (2011): 84–89.

265 *Recent research shows that the brains*: "Patients with OCD Have Difficulty Learning When a Stimulus Is Safe," University of Cambridge, March 6, 2017, https://www.cam.ac.uk/research/news/patients-with-ocd-have-difficulty-learning-when-a-stimulus-is-safe.

268 *As biographer Joshua Wolf Shenk wrote*: J. W. Shenk, *Lincoln's Melancholy: How Depression Challenged a President and Fueled His Greatness* (Boston: Houghton Mifflin Harcourt, Kindle Edition, 2005), 179.

INDEX

Aarhus University, 196
Abbey, Tom, 194–96
Abusive leadership style, 229–31, 238–40
   Bryant's version, 23–28, 31, 34, 124, 174
   Knight's version, 3–5, 9, 12, 15, 22
Acceptance, 70, 174–75, 184–85, 226, 252, 268
Adams State University, 78
Adaptive flexible strategy, 217–20, 222
Adrenaline, 47–49, 122
Adrenaline junkies, 45, 196–97
Adrift (Callahan), 141–42
Adversity, 3, 12, 14, 28–29, 34, 35, 96, 194, 226
   coping strategies for. *See* Coping strategies
Affective inertia, 169–70
"Affective primacy," 119
Affirmations, 155–56
Air Force, U.S., SERE program, 30–32, 33
Airplane pilots, 200, 201
   flight instrument panel analogy, 118–19, 122–23, 130–31
Alabama Crimson Tide, 24
Alexithymia, 115–16, 123
Ali, Muhammad, 145
Altered Traits (Davidson), 169
"Always compete," 14
Alzheimer's disease, 98–99
Ambivalent Parent, the, 148
American Thoracic Society, 20

Amnesia, 29
Amygdala, 45–46, 93–94, 133, 165–67, 169, 187–88, 246–47
Anderson-Kaapa, Drevan, 50–53
Anterior cingulate cortex (ACC), 118–19
Antonovsky, Aaron, 261
Anxiety, 36, 44, 107, 123, 124, 128, 137, 158, 211, 246, 262–63
   brain and, 93–94, 103, 167–68, 205, 219
   Capilano Suspension Bridge experiment, 121–22
   confidence and, 63, 84
   course-correcting for, 57
   discomfort and, 15, 35, 36, 38
   family dinners for, 250
   nuance exercises for, 134–36
   reading the signals, 131, 137, 155
   rumination and, 213, 214, 215
   State-Trait Anxiety Inventory, 199
   steady mind and, 166–69, 180, 186
   training to control, 104–5, 106–7, 186, 189, 205
Apathy, 36, 86, 88–91, 98–99, 101, 111–12, 244, 259
Appelbaum, Mark, 217–18
Apple, 231
Appraisal, 34, 46–50, 53, 57. *See also* Reappraisal
   reading the signals, 128–34, 136
   threat of death and, 43–46
Appropriate goals, 54–55

Army, U.S.
  Human Dimension Strategy, 34
  SERE program, 30–32
  stress response and cortisol, 92–93
Arnold, Katie, 210–12
Aron, Arthur, 121–23, 185
Arousal control, 32, 122–23, 129–30
Arrogance and insecurity, 72–77
Artists, 58, 164, 254
Assembly line approach to teaching, 99–100
Association, 199–200, 204, 217, 218, 222
Ataraxia, 191
Attention control, 32. *See also* Zooming in; Zooming out
  clutch state, 224–25
  emotion regulation and, 205–6, 212–13, 219
  exercises, 182–83
Auschwitz concentration camp, 90–91
Authenticity, 54–55, 247
Authoritarian control, 10–11, 15, 34, 87
  Knight's version of toughness, 3–5, 9, 12, 15, 22
  parenting style, 6–7, 10–11, 12
Autonomous motivation, 234–37
Autonomy, 14, 88, 91, 99, 103, 104, 236
  science of choice and, 95–99
Autonomy-supportive environments, 240–43
Avoidance, 60, 72, 174–75, 181, 187, 262–63, 264
Avoidance learning, 85–87, 96–98
Awareness, 121, 146–47
  calm conversations and, 179–81
  creating space for. *See* Creating space
  of feelings and sensations, 116–19, 132, 134
  of inner voices. *See* Inner voice

Bandura, Albert, 96
Bar, Moshe, 205–6, 207
Baror, Shira, 205–6, 207
Barraza, Brian, 78, 222–23
Baseball, 8, 26, 108
Basic needs, 235–40
Basketball, 9–10, 81–82, 238–39, 241–42, 245–49
  Knight's coaching style, 3–5, 9, 12, 22
Baumrind, Diana, 5–7
Bawden, Mark, 62–63
Beauchamp, Mark, 238
Beckham, Odell, Jr., 73
Bedtime ritual, 263–64
Belonging, 245–50, 251, 261
Bennett, Bonnie, 155
Bereavement, 210–13, 215–16
Bhikkhu Bodhi, 191
Boredom, 177, 181, 182, 248
Boring, Joe, 26
Boston Marathon, 198
Boston Red Sox, 108
Bottom-up processing, 205–6
Bourne, Peter, 92–93
Bowden, Don, 16–17
Brain
  calm conversations and, 187–88
  choice and, 96–99
  control and, 93–94, 98–99, 121, 133
  emotional processing and, 123–24, 127
  fatigue and, 257
  interoceptive system and, 118–19
  modular mind and, 142–45
  OCD and, 265–66
  order and, 38
  pain and meditation, 164–67
  speech and, 148–49
  stress and, 97–98, 246–47
  threat of death and, 45–47, 49
  types of processing, 205–6
Brashness, 73–74, 267

INDEX

Broaden-and-build theory of emotions, 204
Broussard, Fred, 26
Bryant, Paul "Bear," 23–28, 31, 34, 174, 237
Buddhism, 164, 170–71, 191
Bulldoze method, 173, 174, 222
Bullying, 4–5, 8, 12–13
Burnout, 11, 22, 87–88, 167, 168, 194, 239, 255
Burrell, Leroy, 51
Business. *See* Work

California Golden Bears, 16–19
California Task Force to Promote Self-Esteem and Personal and Social Responsibility, 67–69
Callahan, Steven, 139–42
Callousness, 7–9, 12
    downfall of callous view of toughness, 9–12
Calm-and-connect system, 246–47
Calm conversations, 173–91, 262
    applying to your situation, 187–91
    creating space, 176–83
    keeping your mind steady, 183–87
    practicing, 187, 189
Calm Optimist, the, 148
Cameron, James, 124–26
Canadian Athletics Coaching Center, 102–3
Capilano Suspension Bridge experiment, 121–23
Capsaicin, 168
Carleton, Erica, 238
Carroll, Pete, 14–15
Carruthers, Peter, 149
Cavanagh, Peter, 198–99
Central Intelligence Agency (CIA), interrogation techniques, 90
Children. *See also* Parenting styles
    assembly line approach to teaching, 99–100
    confidence and, 62, 65
    confidence and reading levels, 76
    creating space and clip charts, 171–72
    emotion regulation and, 172, 218–20
    language acquisition, 152–53
    pacing and fitness test, 41–42, 256
    perseverance and, 156–57
    self-esteem and, 68–69
Chiu, Ming Ming, 75–77
Choice
    give-up-itis, 88–91
    giving yourself, 105–6, 112
    science of, 95–99
Christianity, 191–92
Classical conditioning, 85–86
Cleather, Dan, 163–64
Clip charts, 171–72
Clutch (clutch state), 168, 224–26
Coaching, 14–16
    Bryant and the Junction Boys, 23–28, 31, 34, 124, 174
    Carroll's style of, 14–15
    control and failure, 111
    control and leading others, 109–11
    control and training, 100–101
    Julie and cults, 229–30
    Knight's style of, 3–5, 9, 12, 22
    leadership styles, 15, 229–31, 237–40
    making progress and growing, 243–45
    McNair's style of, 9–10
    need to belong and, 245–50
    power and, 102–3
    supporting not thwarting, 240–43
    Wilt and Edelen, 59–61, 81
Cognitive biases, 80
Cognitive development, 152–53, 154
Cognitive narrowing, 200–201

Cognitive strategies, 34–35, 213, 219–20, 221
Cognitive zooming, 209
Comfort zone, 159, 244
Compartmentalization, 32, 83–84, 214
Compassion, 22, 148–49
Compensatory control, 109
Competence, 236, 244
Comprehensibility, 261
Comprehensive Soldier Fitness (CSF), 33
Concentration, 209, 212–13
Concentration camps, 90–92, 255, 258–61, 263
Conditional regard, 231
Conditioning, 9–10, 12, 100–101
Confidence, 59–84
  building wrong kind of, 66–69
  contingent self-worth and seeking self-esteem, 69–72
  creating inner. *See* Creating inner confidence
  Instagram version of, 62
  old model of, 62
  role of experience in, 72–77
  shaping how you see the world, 62–65
  Wilt and Edelen, 59–61, 81
Confrontational dialogue, 147–48
Conscious awareness, 119, 121, 146–47, 180–81
Constraints, setting and letting go, 87, 99, 110–11
Contingent self-worth, 69–72
Control
  brain and, 93–94, 98–99, 121, 133
  choosing to be tough, 92–95
  coaching and training, 100–101
  leading others, 109–11
  leading yourself, exercises, 104–9
  training yourself. *See* Training to control
  VO_2 max test, 94–95, 101

Controlling leadership style, 229–31, 238–40
  Bryant's version, 23–28, 31, 34, 124, 174
  Knight's version, 3–5, 9, 12, 15, 22
Conversation
  calm. *See* Calm conversations
  creating connections, 247–49
  inner dialogue. *See* Inner voice
Cooper, Kenneth, 198–99
Cooperation, 245–46
Coping statements, 154–55
Coping strategies
  adjusting state of mind, 204–10
  attending to discomfort, 197–200
  capacity to cope, 220–23
  exercises changing processing ratio, 208–10
  exercises letting mind to go bad place, 190–91
  for fear, 194–97
  flexible and adaptable, 217–20
  letting mind go to a bad place while performing, 190–91
  suppressing vs. addressing, 210–16
  zooming in vs. out, 190, 200–204
Cortisol, 47–49, 92, 246–47
Costill, David, 198–99
Courage to Create, The (May), 172–73
Course-correcting for stress, 56–57
Covey, Stephen, 171
COVID-19 pandemic, 46, 231–32
Coyle, Dan, 250
Creating inner confidence, 77–83
  developing quiet ego, 82–83
  lowering the bar and raising the floor, 78–79
  shedding perfection, 79–80
  trusting training and yourself, 81–82
Creating space, 13, 16, 159, 171–72, 176–83, 192

exercises for, 182–83
spending time alone in your head, 176–83
Creativity, 202–3, 210
Crow, John David, 25
Csikszentmihalyi, Mihaly, 223, 225
Cult-style environment, 229–31, 238–40
Culture Code, The (Coyle), 250

Damasio, Antonio, 119, 123–24
D'Antoni, Mike, 14–15
Darwinism, 27, 117
Davidson, Richard, 169, 170
Davis, Hap, 187–88
Daydreaming, 225
Death, fear of, 43–46
Death of loved one and grief, 210–13, 215–16
De Botton, Alain, 84, 160
Deci, Edward, 235–36, 240, 251
Decision-making
modular mind and brain, 142–45, 146
role of emotions in, 123–24, 131, 136–37
"trolley problems," 126–27
Defining success, 55–56
Dehydration, 140
Deliberate rumination, 214, 262
Delusion, 69, 73, 79
Demandingness, 5–7, 9, 11, 13, 16, 34, 237
Denison, Jim, 102–3
Dent, Jim, 24–25
Depersonalization, 29
Depression, 98, 132, 133, 166, 203, 214, 250
Depression Adjective Checklist, 199
Derealization, 29
Descartes' Error (Damasio), 123–24
Deschamps, Thibault, 56

Despair, 63, 88–89, 212, 215, 233, 259–61, 263
Detachment, 29, 53, 169, 213, 216. *See also* Dissociation
Detention, in SERE training, 30–31
Deukmejian, George, 67
"Devil on our shoulder," 63–64
Discipline, 3, 4, 6–7, 11–14, 87–88, 231
Discomfort
attending to, 197–200
author's experience of, 17, 19–21
brain and pain, 164–67
building the foundation to do hard things, 229–52
choosing the difficult path, 37–38
finding meaning in. *See* Meaning in discomfort
military model of toughness, 35–36
redefining toughness, 12, 13, 15–16
steady mind for, 184–87, 188
turning into action, 36–38
Discovery (ship), 88
Disgust, 45, 117, 123–24, 126, 127–28, 144, 259
Dissociation, 29–30, 199–200, 204, 217–18, 233
Distractions, 36, 218, 221, 222, 225
brain and, 167
calm conversations and steady mind, 177–83
children and, 156–58, 219
death of loved one and grief, 212–13, 215, 216
exercises for, 182–83, 210
zooming out, 201–2, 210
Doctor and the Soul, The (Frankl), 258
Don Bowden Mile, 16–19
Dorsal raphe nucleus (DRN), 97–98
Doubts, 17, 21, 43, 60–64, 74
Dread, 47, 94, 104, 107, 128, 166, 175
Drive, 256–58

Drug abuse, 10–11, 67–68, 130, 207, 236, 250
Dudley, Ed, 25
Dungy, Tony, 14–15
Dutton, Donald, 121–23

Eastern Washington University, 242
Eating disorders, 130, 250
Edelen, Leonard "Buddy," 59–61, 81
Ego, 79–80, 82–83, 112
Ego vs. Soul in Sports (Reed), 14–15
El Capitan (Yosemite National Park), 44–46
Embracing reality, 50–58, 79–80, 169
  setting mind on reality, 53–58
Emory University, 92
Emotions, 15, 63, 64, 115–38. *See also* Feelings; *and specific emotions*
  alexithymia, 115–16, 123
  appraisal and reading the signals, 128–34
  awareness of feelings and sensations, 116–19, 132, 134, 199–200
  exercises for developing nuance, 134–37
  feelings vs., 116–17
  interpretation and contextualization of, 120–24, 132
  listening to, 120–24
  power of what we feel, 116–19
  role in decision-making, 123–24, 131, 136–37
  sending a message, 124–28
  subselves and modular mind, 144–45
  use of term, 117
Emotional regulation, 136–37, 167, 180, 197
  adjusting state of mind, 204–10
  attending to discomfort, 197–200
  brain and, 98–99
  capacity to cope, 220–23
  coping with fear, 194–97
  decision making and, 136–37
  exercises changing processing ratio, 208–10
  flexible and adaptable, 217–20
  process model of, 220–21
  suppressing vs. addressing, 210–16
  use of term, 197
  zooming in vs. out, 200–204
"Emotion-Regulation Choice" (Sheppes and Gross), 213–14, 216
Emotion wheel, 133, 135
Empathy, 148–49
Endurance
  sensation of effort and, 41–43, 256
  VO_2 max test, 94–95, 101
  Willie (dog), 255–56
English Institute of Sport, 62–63
Environmental zooming, 210
Equanimity, 191–92, 226
Evaluative integration, 83–84
Evolution, 57–58
Exercises
  changing processing ratio, 208–10
  creating and amplifying using imagery, 183
  creating space, 182–83
  developing nuance, 134–37
  leading others, 109–11
  leading yourself, 104–9
  letting mind go to a bad place while performing, 190–91
  mistake watching, 189–90
  noticing, 182
  turning the dial, 182–83
Exercising and listening to music, 176–78
Expectations, 74–75
  appraisal of situation and, 46–50
  defining, 55–56
  reframing, 78–79

setting goals. *See* Goals
setting mind on reality, 53–58
threat of death and, 43–46
"Experimentum Crucis," 253–55
Expressive flexibility, 215–16
External regulation, 71–72
Extrinsic motivation, 71–72
Eysenck Personality Inventory, 199

Facade, 12, 22, 51, 79
Facing reality. *See* Embracing reality
"Fainting" rooms, 143
Faithful Friend, the, 148
Fake confidence. *See* False confidence
"Fake it until you make it!", 65, 74, 77, 156
Fake toughness, 12–13, 15, 73–74
False confidence, 9, 43, 69, 72–77, 79–80, 267
Family dinners, 250
Family Feud style of thinking, 209
Fatigue, 9, 15
   adaptive flexible strategy for, 217–18
   course-correcting for, 56–57
   inner voice and, 141
   packing and sensation of effort, 42–43
   runner tells, 17–19
   steady mind and, 168, 173, 175, 176
   will to continue, 256–58
Fear, 144–45, 252
   authoritarian style and, 5, 6–7, 11, 12–13, 237, 240, 242, 244
   Capilano Suspension Bridge experiment, 121–22
   coping with, 186, 194–97
   from fear to despair to apathy, 258–63
   inner voice and, 141, 144–45, 149
   Knight's version of toughness, 3–5, 9, 12, 15, 22
   letting go, 109–10

nuance of inner strength, 263–66
risk taking and, 244–45
threat of death, 43–46
training learned hopefulness and, 101–2
trusting your training, 81–82
Feder, Adriana, 262
Feelings, 116–19. *See also* Emotions; *and specific feelings*
   awareness of, 116–19, 132, 199–200
   exercises for developing nuance, 134–37
   interpretation and contextualization of, 120–24, 132
   messengers vs. dictators, 136–37
   as predictive, 120–24
   sending a message, 124–28
   subselves and modular mind, 144–45
Fernyhough, Charles, 149–50
Fighting through, 174–75, 187
Fight-or-flight response, 47–48, 49, 246–47
First-person pronouns, 158, 207, 210
First pillar of toughness. *See* Pillar 1
Fist bumps, 245–46, 248
Flexibility, 14, 30, 34, 203, 215–20, 222
Flipping the script, 106–8
Flow, 223–25
Flow Genome Project, 224
Flu vaccine, 32–33
Football, 3, 9–10, 14, 72–73
   Bryant and the Junction Boys, 23–28, 31, 34, 174
Forbes (magazine), 231
Ford, Henry, 99
Foucault, Michel, 102
Four-minute-mile, 16–19
Four pillars of real toughness, 22, 41–268
   first pillar, 22, 41–112
   second pillar, 22, 115–60

third pillar, 22, 163–92
fourth pillar, 22, 229–68
Frankl, Viktor, 90–91, 171, 255, 258–61, 263
Frazier, Joe, 145
Freak-outs, 37, 42, 45, 124, 137
    author's experience with, 18–19, 20, 176
    confidence and, 64, 76
    setting mind on reality, 54, 57, 58
    steady mind and, 165, 166, 176, 190–91
Fredrickson, Barbara, 204
Freedom, 172, 260, 263
Freeman, Mark, 70–71
Free soloing, 44–46
Fuqua, Theresa, 19

Garciaparra, Nomar, 108
George Mason University, 262
Girl Scout Cookies, 215
Give-up-itis, 88–91
"Give-Up-Itis Revisited" (Leach), 90–91
Goals (goal setting), 31, 32, 54–55, 75, 78, 190, 234
Godspeed (ship), 88
Goehring, Dennis, 24
Golden State Warriors, 241–42
Goldilocks parenting style, 6–7
Gonzales, Britani, 178–80
"Good helpers," 156–57
Google, 245
Grade inflation, 72
Green, Draymond, 241–42
Greene, Joshua, 127
Greene, Roberta, 261
Grief, 210–13, 215–16
"Grind," 8, 15, 18–19, 61, 70, 91, 101
Gross, James, 213–14, 216
Growth. *See* Post-traumatic growth

Habits, 105–6, 235
Haidt, Jonathan, 127–28
Haude Elementary School, 41–42
Haunted house visitors, 195–96
Hayes, Steven, 154–55
Hays, Kate, 62–63
Hearing Voices Movement, 149–50
Heart rate, 59, 121, 122, 129, 132–33
Helpless Child, the, 148
Helplessness. *See* Learned helplessness
Herskowitz, Mickey, 24
Herz, Noa, 205–6, 207
Hierarchy of needs, 251–52
Highlen, Pamela, 155
Hinduism, 191
Ho Chi Minh Trail, 92
Holistic health and fitness, 34
Holmes, Baxter, 248
Holocaust, 90–91, 258–61
Homeostasis, 116
Honnold, Alex, 44–46
Hopefulness. *See* Learned hopefulness
Hopelessness, 87–88, 97–99
Houston Cougars, 16–19, 50–51, 179
Houston Post, 24
How to Change (Milkman), 105
Huang, Cassey, 245
Humility, 15, 74
Hyman, Mark, 8

Ignoring, 121, 168, 174–75, 222
Iguodala, Andre, 241–42
Imagery, 32, 183
Imagine the future, 209–10
Immune system, 32–33, 48
Inattentional deafness, 200–201, 207
Indiana Hoosiers, 4
Inner confidence, 13, 76–83. *See also* Creating inner confidence

INDEX

Inner drive, 232–34
Inner strength, 263–66
Inner voice, 111–12, 146–60, 203
   author's experience of, 18–19
   Callahan's split characters, 139–42
   changing your, 152–55
   choosing the difficult path, 36–38
   creating a ritual, 109
   decreasing the bond, 156–59
   Honnold and El Capitan, 44–45
   knowing which to listen to, 155–56
   modular mind and subselves, 142–45
   quitting and, 35–36, 149
   winning the inner debate. *See* Winning the inner debate
"Inner warrior," 9
Insecurities, 12–13, 14, 43, 60, 61, 62–64, 79–80
   arrogance and, 72–77
Inside Out (movie), 144–45
Instagram, 62, 267
Integrated dialogue, 147–48
Interoception, 118–19, 123, 129–32, 138, 179–81
Interpersonal closeness, 184–85
Intimidation, 231
Intrinsic motivation, 235–40
Introspection, 55
Intrusive thoughts, 146–47, 264–66
Iraq War, 248
Isolation, 176–78, 180, 231
Iyengar, Sheena, 99

James, LeBron, 81
Jamestown Colony, 88–89
Jerabek, Ilona, 76–77
Jobs. *See* Work
Jobs, Steve, 231
Job satisfaction, 96, 232
Jogging, 59–60

Johnson, Lyndon B., 41
Jordan, Michael, 81–82
Joseph, Jane, 45–46
Joseph, Moise, 194–96
Journaling, 55, 210
Judgments, defining, 55–56
Junction Boys, The (Dent), 24–25

Karate Kid (movie), 8
Kashdan, Todd, 262
Kaufman, Scott Barry, 192, 226, 246
Keith, Bobby Drake, 27
Keltner, Dacher, 245
Kentucky Wildcats, 23
Kerr, Steve, 241–42
Kettler, Elwood, 23–24
Kimmel, Jimmy, 125
Kipling, Rudyard, 30
Kitty Hawk, USS, 26
Klein, Ezra, 170
Knight, Bobby, 3–5, 9, 12, 22, 237
Korean War, 30, 89
Kotler, Steven, 223–24
Kraus, Michael, 245
Kross, Ethan, 159
Krueger, Charlie, 25
Kupperman, Karen, 89

Labeling, 135, 190
Ladders, 43–44
Language acquisition, 152–53
Lapate, Regina, 180–81
Late positive potential (LPP), 219
Leach, John, 90–91
Leadership styles, 15, 229–31, 237–40. *See also* Coaching
Leading others, 109–11
Leading yourself, exercises for. *See* Training to control
Leadville Trail 100 Run, 212

League of Their Own, A (movie), 8
Learned helplessness, 87, 90, 96–98, 101–2
Learned hopefulness, 98–104
   training, 99–104. *See also* Training to control
Leotti, Lauren, 99
Letting go, 109–10
Lewis, Carl, 51
Life satisfaction, 69–70, 96
Linguistic zooming, 158, 210
Listlessness, 86–87, 89–90, 91, 98
Love, 185, 192
Lower the bar, 78–79
Lutz, Antoine, 164–65

Machismo, 8–9, 34, 130
Mackey, Danny, 184
McNair, Jordan, 9–10
MacTavish, Craig, 3
Magness, Bill, 66, 264
Magness, Elizabeth, 66, 264
Mahoney, John, 236–37
Maier, Steven, 85–87, 96–99
Manageability and coherence, 261
Man's Search for Meaning (Frankl), 258
Marathons, 42–43, 48, 59–60, 197–99, 201, 216, 217–18
Marie-Galante, 140–41
Maryland Terrapins, 9–10
Masculinity, 8–9
Masking, 51, 79–80
Maslow, Abraham, hierarchy of needs, 251–52
May, Rollo, 172–73
Meaningful life, 253–54
Meaningfulness and coherence, 261
Meaning in discomfort, 253–68
   from fear to despair to apathy, 258–63
   nuance of inner strength, 263–66
   will to continue, 255–58

Meditation, 13, 164–73, 176–78, 180
   pain and monks, 164–67
Meggs, Jennifer, 83–84
Mental imagery, 32, 183
Mental skills coaching, 31–32, 33
Methodist Church, 191
Michigan Wolverines, 4
Micromanaging, 87, 96, 109–10
Mile run, 16–19, 41–42
Miles, Stephen, 19
Military, 28–36
   authoritarian parenting and, 11
   control and failure, 111
   emotional bond between soldiers, 248
   model of toughness, 11, 14, 16, 28–36
   stress response and cortisol, 92–93
   survival training, 30–32, 52–53
Military Academy, U.S., Center for Enhanced Performance, 31
Milkman, Katy, 105
Millennials, 69
Mills, Joseph, 102–3
Mind. *See* Modular mind; State of mind; Steady mind
Mindfulness, 171, 174, 179–80
Misattribution of arousal, 122–23, 129–30
Mistake watching, 188–90
   exercises, 189–90
Mnuchin, Steve, 231–32
Modular mind, 142–45, 146
Monks and meditation, 164–67, 192
Montgomery, Hillary, 156–57, 171–72, 255
"Mood follows action," 206, 207, 209
Moore, Kenny, 198–99
Moral reasoning, 126–28
Morgan, William, 199, 217–18
Mosaic, 115
Motivation, 6, 11, 14, 54, 71–72, 256–58

INDEX

autonomous, 234–37
   filling basic needs, 235–40
Mount Everest, 233
Mount Sinai School of Medicine, 262
Music, 176–78, 182

Nadal, Rafael, 108
Naming emotions, 135, 190
Narcissism, 69
Narrow focus, 200–204, 206–8
Nature (magazine), 159
Nautilus (magazine), 45
Navy, U.S., 26, 30–32
Navy SEALs, 28–29, 32
NBA (National Basketball Association), 238–39, 241–42, 245–49
Negative bias and priming, 57–58
Negative conditional regard, 231
Negative self-talk, 155–56, 159–60, 174
Negative thoughts and feelings, 63–64, 83–84, 101, 137–38, 204
   brain and control, 98–99, 148
   quitting and, 35–36
   steady mind and, 182, 186, 187
Neuroscience. *See* Brain
Newton, Isaac, 253
New York City Marathon, 198
New York Times, 101
Nolan, Emily, 184
Nolan, Katie, 72–73
Noticing, 182
Nuance
   exercises for, 134–37
   of inner strength, 263–66
Nursing home residents, 96

Obsessive-compulsive disorder (OCD), 263–66
Obsessiveness, 60
Ochsner, Kevin, 99

Ohio University, 10
Olympic Games (1972), 197–98
Olympic Games (2016), 19*n*
Olympic Trials (1964), 61
Olympic Trials (1988), 217–18
On Confidence (de Botton), 84
Oprah Winfrey Show, The, 67, 68
Outside (magazine), 211
Overconfidence. *See* False confidence
Oxytocin, 47–48, 246–47

Pace (pacing), 17–18, 94, 199–200, 256
   children and fitness test, 41–42, 256
   sensation of effort and, 41–43
Pain, 3, 163–73
   attending to, 225–26
   author's experience of, 19–21, 173–75
   brain and, 96–99, 164–69, 246
   Bryant and the Junction Boys, 23–28, 31, 34, 124, 174
   calm conversation for, 173–79, 191
   Cleather and tattoos, 163–64
   death of loved one and, 211–13
   meditation for, 164–71, 176–78
Pan American Games, 155
Panorama mode, 208–9
Pardee, Jack, 27
Parenting styles, 5–7, 10–11
Pavlov, Ivan, 85
"Paying our dues," 69
Peak Performance (Magness and Stulberg), 58
Perfectionism, 6, 22, 43, 61, 79–80
Performance state, 47–48, 224
Perseverance, 3, 156–57
Persistence, 43, 64, 94, 95, 112, 232–34, 258
"Phantom vibration syndrome," 120
Phoenix Suns, 241
Phones, 36, 120, 181, 182, 186, 201, 250

Physical zooming, 209
Pillar 1 (ditching the facade and embracing reality), 22, 41–112
  accepting what you are capable of, 41–58
  knowing when to hold 'em and when to fold 'em, 85–115
  true confidence, 59–84
Pillar 2 (listening to your body), 22, 115–60
  emotions as messengers, 115–38
  owning the voice in your head, 139–60
Pillar 3 (responding instead of reacting), 22, 163–92
  keeping your mind steady, 163–92
  turning the dial so you don't spiral, 193–226
Pillar 4 (transcending discomfort), 22, 229–68
  building the foundation to do hard things, 229–52
  finding meaning in discomfort, 253–68
Pineda, Nate, 72
Pixar, 144–45
Pocahontas, 89
Pollock, Michael, 199, 217–18
Popovich, Gregg, 247, 248–49
Portrait mode, 208–9
Positive self-talk, 155–56, 159–60, 186–87
Post-traumatic growth (PTG), 214, 261–63
Post-traumatic stress disorder (PTSD), 116, 261
Power, 102–3
POWs (prisoners of war), 30, 89, 261–62
Praise (praising), 66–67, 70–72
Prefontaine, Steve, 198–99
Prefrontal cortex (PFC), 94, 97–99, 118–19, 123, 133, 166–68
Presidential Physical Fitness Test, 41–42

Priming, 57–58, 64, 206
Process model of emotion regulation, 220–21
Profile of Mood States, 199
Progression, 243–45
Prot-Klinger, Katarzyna, 261
Proud Rival, the, 148
Psychological distance, 158
Psychological safety, 244–45, 251
"Psychology of the Concentration Camp, The" (Frankl), 255
Puchalska-Wasyl, Małgorzata, 147–48
Purpose, 256–58

Quiet ego, 82–83
"quiet the mind," 177, 212
Quitting, 35–36, 105–6, 149
  Bryant and the Junction Boys, 23–28, 31, 34

Raise the floor, 78–79
RAND Corporation, 32
Reactivity, 120–21, 183–87
Reading the signals, 128–34
Reality, embracing. *See* Embracing reality
Reappraisal, 136, 174, 213–16, 218, 219, 221, 222
Reed, Ken, 14
Reengagement, 17, 233–34
Reframing, 78–79, 190
Rejection, 159, 202, 246
Relatedness, 236
Relationships and arguments, 193–94, 201
Resets, 172
Resilience, 13, 16, 30, 31, 33, 131–32, 242, 267
Responsiveness and parenting styles, 5–7
Rewards, 70–72, 96, 231
Rhabdomyolysis, 10
Ridpath, B. David, 10

INDEX

Righteous Mind, The (Haidt), 127–28
Risk taking, 244–45
Rituals, 108–9
Rock climbing, 44–46
Roll, Rich, 22, 207
Rumination, 83, 150, 158, 178, 203–4, 206, 212–13, 214, 262
Runners (running). *See also* Marathons
   adaptive flexible strategy for, 217–20, 222
   Anderson-Kaapa and facing reality, 50–53
   Arnold and grief, 211–12
   attending to discomfort, 197–200
   author's experience, 16–21, 41–42, 106–7, 153–53
   four-minute-mile, 16–19
   Gonzales and, 178–80
   power and, 102–3
   role of toughness for, 21–22
   VO_2 max test for, 94–95
   Wilt and Edelen, 59–61, 81
Runner tells, 17–19
Running Home (Arnold), 211–12
Ryan, Richard, 235–40, 251

Sadness, 117, 133, 136, 144, 145, 213
Salary and job satisfaction, 232
San Antonio Spurs, 247
Satisfaction, 69–70, 96, 232
Scientific American, 82
Second-person pronouns, 158, 159, 207, 210
Second pillar of toughness. *See* Pillar 2
Self-actualization, 226, 251
Self-awareness. *See* Awareness
Self-belief, 64–65. *See also* Confidence
Self-confidence. *See* Confidence
Self-determination theory (SDT), 95–99, 236–37, 240, 251

Self-distancing, 158–59
Self-doubt. *See* Doubts
Self-efficacy, 95–99
Self-esteem, 66–72, 80, 156, 251. *See also* Confidence
   seeking, 69–72
Self-esteem movement, 66–70
Self-harm, 128–30
Self-immersion, 158–59
Self-motivation. *See* Motivation
Self-protection, 144–45, 191
Self-talk, 31, 32, 147–50, 155–56, 159–60, 186–87, 262. *See also* Inner voice
Self-transcendence, 251–52
Self-worth, 69–72
Seligman, Martin, 85–87
Sensations, 18, 37–38, 103, 120–21, 131
   awareness of, 116–19, 132, 134
   of effort and endurance, 41–43
   feelings as predictive, 120–24
   perception and, 205–6
   role of interoceptive system, 118–19
Sense of purpose, 256–58
Sensory deprivation tanks, 180
SERE (Survival, Evasion, Resistance, and Escape) program, 30–32
Setting mind on reality, 53–58
Sharot, Tali, 57
Sheppes, Gal, 213–14, 216
Shorter, Frank, 197–99, 204, 208
Shula, Don, 14–15
Silva, John, 217–18
Sink or swim, 34–35
Skydivers, 47, 48
Sleep as basic need, 251
Sleep deprivation, 52, 56
Smartphones, 36, 120, 181, 182, 186, 201, 250
Smelser, Neil, 67–68

Smith, Dean, 14–15
Smith, John, 89, 91, 93
Snowball on the hill, 220–21
Snow Farm, 183–85
Social recovery, 247
Sociometer theory, 70
"Soft," 4–5, 7, 8, 14
Solitary confinement, 176–78
Solomon, Richard, 85–86
Somatic Perception Questionnaire, 199
Soreness-versus-injury, 130–31
Sorensen, Meredith, 106–7
Specifity and nuance, 134
Speech and brain, 148–49
Spiller, Rob Roy, 24
Sports Illustrated, 60, 198
Springfield College, 156
Stallings, Gene, 24–25
Stanford University, 213–14
State of mind, 204–10
State-Trait Anxiety Inventory, 199
State University of New York (SUNY), Buffalo, 75–76
Steady mind, 163–92
   calm conversations. *See* Calm conversations
   Cleather and tattoos, 163–64
   developing ability to respond instead of react, 183–87
   equanimity, 191–92
   pain and meditation, 164–73
   spending time alone in your head, 176–83
Stevens, Brad, 14–15
Stimulus and response, 166–71
Stock traders, 131
Stoicism, 3, 9, 18, 29, 140, 191
Storr, Will, 68
Stover, Andy, 106–7
Strachey, William, 88–89

Strange Order of Things, The (Damasio), 119
Stress, 28–29, 208
   appraisal of situation and, 46–49
   course-correcting for, 56–57
   level of control and, 92–95
   military model and, 28–29, 34–35, 52–53, 92–93
   priming the mind, 57–58
   resilience and, 131–32
   Trier Social Stress Test (TSST), 168–69
Stress inoculation, 31–33, 34–35
Stress response, 45–49, 54, 92–93, 96–98, 166–69
Subselves, 143–45, 146–47, 160
   Callahan's split characters, 139–42
Substance abuse, 10–11, 67–68, 130, 207, 236, 250
Suffering, 3, 13, 87, 104, 218, 254
Suicidal thinking, 116
Supportive environments, 240–43
Suppression, 15, 124, 136, 174–75, 192, 196, 213, 214–15, 222
Survival-of-the-fittest, 27–28
Survival training, 28–34, 52–53
Survive, will to, 31, 260–62
Susan Constant (ship), 88
Swann, Christian, 224, 225, 226
Swansea University, 129

Talladega Nights (movie), 55
Tantrums, 4, 12, 130, 166, 172, 193, 194, 219
Tattoos, 163–64
Teague, Foster "Tooter," 26
Team dinners, 247, 249, 250
Teesside University, 83–84
Temporal zooming, 209–10
Testosterone, 49
Test scores, 6–7, 100

# INDEX

Texas A&M Aggies, 23–28, 31, 34
Texas Tech Red Raiders, 25
Theresienstadt, 258–61
Third-person pronouns, 158, 159, 207, 210
Thorpe, George, 89
Threat of death, 43–46
Thrill-seekers, 196–97
Thwarting autonomy, 242–43
Time (magazine), 188
Titanic (movie), 124–26
Top-down processing, 205–6
Torrence, David, 19*n*
Toughness
    capacity to cope, 220–23
    choosing the difficult path, 36–38
    choosing to be tough, 92–95
    confidence and, 64–65
    downfall of callous view of, 9–12
    feelings and emotions, 33, 117
    four pillars of real. *See* Four pillars of real toughness
    pain and, 163–64
    parenting styles and, 5–7, 10–11
    redefining, 12–16, 38
    searching for real, 16–22
    as a society, 267–68
Toughness, models of
    of Bobby Knight, 3–5, 9, 12, 22
    Bryant and the Junction Boys, 23–28, 31, 34, 124, 174
    of military, 11, 16, 28–36
    old-school, 8–9, 23–28, 34, 64–65, 87, 116, 142, 149, 194, 206, 222, 237–38, 267–68
Toughness maxims
    appraisals of situation, 50
    confidence as a filter, 65
    control and capacity to cope, 104
    embracing reality, 53
    exploring instead of avoiding, 263
    feelings send a message, 128
    feelings subject to distortion, 124
    flexible and adaptive coping ability, 216
    inner speech, 150
    level of control and stress, 95
    real toughness, 16
    responding to stress, 173
    satisfying basic needs, 237
    sense of purpose, 258
Toward a State of Esteem (Vasconcellos), 67–68
Training
    learned hopefulness and, 99–104
    power and, 102–3
    trusting your, 81–82
    Wilt and Edelen, 59–61, 81
Training to control, 104–9
    adopting rituals, 108–9
    flipping the script, 106–8
    giving yourself a choice, 105–6
    small to large, 104–5
Transcend (Kaufman), 192, 226, 246
Transcending discomfort, 22, 229–68
    building the foundation to do hard things, 229–52
    finding meaning in discomfort, 253–68
Trier Social Stress Test (TSST), 168–69
"Trolley problems," 126–27
Trump, Donald, 73–74
Trust, 15, 77, 110, 240, 245–46, 249
    training, 81–82
"Trust but verify," 110
Twenge, Jean, 68–69

Unemployment insurance, 231–32
University College London, 57
University of Arkansas, 36
University of California, Los Angeles, 133
University of California, San Diego, 131

University of Lublin, 147–48
University of Maryland, 202–3
University of Michigan, 156, 158, 159, 201
University of Nantes, 56
University of Pennsylvania, 85, 126, 156
University of Waterloo, 156
University of Wisconsin, 168–69, 180
University of Wisconsin, 164
Until It Hurts (Hyman), 8
Upekkha, 191

Vaccine analogy, 31, 32–33
Van Raalte, Judy, 156
Vasconcellos, John, 67–68
Ventromedial prefrontal cortex (vmPFC), 123
Vietnam War, 92–93, 261–62
Visualizations, 106, 183
Visual zooming, 208–9
Vocabulary, naming emotions, 135
Vocal cord dysfunction (VCD), 19–21, 173–74
Voice in your head. *See* Inner voice
Voices Within, The (Fernyhough), 149–50
VO_2 max test, 94–95, 101
Vulnerability, 8, 74, 79–80
Vygotsky, Lev, 152–53

Wai, Jonathan, 36
Walsh, Bill, 14
Washington Post, 24
Waterboarding, 90
Wayment, Heidi, 82–83
Wesley, John, 191
"What if?" thoughts, 38, 44
White, Rachel, 156–57
White-knucklers, 196–97
Why Buddhism Is True (Wright), 117, 145, 146–47
Williams, Serena, 73, 108

Willie (dog), 255–56
Will to continue, 255–58
Will to survive, 31, 260–61
Wilson, Timothy, 177–78
Wilt, Fred, 59–61, 81
Winning the inner debate, 151–59
  changing your voice, 152–55
  decreasing the bond, 156–59
  knowing what voice to listen to, 155–56
Winslet, Kate, 125
Wooden, John, 14–15
Work (workplace), 21, 87, 99, 101, 242
  false confidence and, 77
  give-up-itis and, 91
  importance of social interactions, 249–50
  job satisfaction, 96, 232
  risk taking and safety, 244–45
  role of constraints, 110–11
Work ethic, 4, 70, 232, 235
"Work or starve," 89
World University Games, 50
Wright, Jim, 25
Wright, Robert, 117, 145, 146–47
Wundt, Wilhelm, 119

Yale University, 166–67
Yes to Life (Frankl), 260
Yoga, 13, 166, 170, 191
You Are Not a Rock (Freeman), 71
Young, Hayley, 129

"Zone out," 29, 53, 199, 217
Zooming in, 190, 200–202, 206–10, 222
  exercises, 208–9
Zooming out, 190, 201–4, 206–10, 222
  exercises, 208–9
Zuleger, Brian, 32, 78